Eyvind Claxton

About the Author

JONATHAN COHN is a senior editor at *The New Republic*, where he has written about national politics and its impact on American communities for the past decade. He is also a contributing editor at *The American Prospect* and a senior fellow at the think tank Demos. Cohn, who has been a media fellow with the Kaiser Family Foundation, has written for *The New York Times*, *The Washington Post*, *Mother Jones*, *Rolling Stone*, *Slate*, and *The Washington Monthly*. A graduate of Harvard, he lives in Ann Arbor, Michigan, with his wife and two children.

"An important book. . . . *Sick* is much more than a meticulously drawn and moving compilation of crises. It is also an edifying primer on how we got here."
—*The New York Times Book Review*

"Jonathan Cohn of *The New Republic* is one of the best health care writers out there."
—David Leonhardt, *The New York Times*

"Cohn offers a brilliant and sympathetic narrative of how the people who try to make our health system work actually do their jobs, and how patients try to navigate a system that is at times wondrous—and at other times impossible."
—E. J. Dionne, Jr., syndicated columnist and author of *Why Americans Hate Politics*

"*Sick* focuses in heartrending detail on nine stories, the kind which may well find their way into stump speeches in 2008. Cohn brings a fresher perspective to the health care debate."
—*Newsweek*

"Jonathan Cohn weaves personal tragedies and policy failures into a tapestry of shame. His book will infuriate you enough to make you want to scream at every member of Congress, 'Read this!'"
—David K. Shipler, author of *The Working Poor*

"Mr. Cohn relates vignette after wrenching vignette of real folks' experiences to illustrate, in human terms, what can happen when insurance—and health care—fails those who fall between the cracks."
—*The Wall Street Journal*

"In *Sick*, Jonathan Cohn takes an honest, penetrating look at a health care system which is completely broken. Cohn has written a call to arms for a complete transformation of American medicine."
—Alex Kotlowitz, author of *There Are No Children Here*

"Vivid and engaging. . . . Highly readable. . . . Mr. Cohn is a gifted storyteller, and he uses his skill to turn a wonkish matter into an uncommonly good read."
—*The New York Sun*

"Who should read *Sick* by Jonathan Cohn? Every political candidate for president. Every insurer. Every person who has ever been hospitalized or will be hospitalized. In other words, everybody."
—Buzz Bissinger, author of *Friday Night Lights*

"*Sick* may be the first book on health care policy that might accurately be described as 'a good read.' It's filled with well-reported individual stories about various medical crises and some nicely summarized background detail on how our current health system evolved."
—*National Review*

"Fascinating. . . . This is a terribly important book. . . . Cohn keeps the reader in suspense by telling stories in two acts, separated by historical details of the development—and failures—of the nation's health care industry."
—*The Chicago Sun-Times*

"Heart-wrenching. . . . Cohn tells his stories with compassion and rich detail, vividly demonstrating how the cost of health care threatens not only the finances of ordinary Americans but, quite literally, their very survival."
—*The Washington Monthly*

"A brilliant new book analyzing our health care system. Cohn tells some remarkable stories of people who are suffering because of the system."
—Tim Johnson, *ABC News*

"A gripping account of our broken health care system. With his characteristic humanity, Cohn explores the struggles of everyday Americans to secure affordable health care and trumpets the call for reform."
—*The New Republic*

"Note to Americans: Read Jonathan Cohn's book for a vivid, and frequently terrifying, account of what fate could await you, your family, and your friends."
—*The Chicago Tribune*

"An eye-opening look at the failures of the nation's system."
—*The Los Angeles Times*

"Cohn offers a uniquely thorough diagnosis of the problems we face. . . . He manages his stories to great effect, ably describing people left without options or hope in their hour of need."
—*Commentary*

"Cohn is a terrific storyteller. . . . An honest read. It's hard not to come away appalled by the American health care system. . . . *Sick* reads like a novel.
—*Columbia Journalism Review*

"Detailed, controlled, comprehensive, analytical, and, above all, intelligent. . . . *Sick* provides just about as accurate an overview of the terrain as being a doctor or a patient does these days. . . . It is a pleasure to have the whole drama laid out, act by act."
—Abigail Zuger, *Health Affairs*

"As a compelling, authoritative analysis of what's wrong, how we got there, and what our system is doing to our fellow citizens, Cohn's eminently readable book will draw even the most jaded observer into a useful state of coherent fury. Think *Arrowsmith* meets *Sicko* meets *The New England Journal of Medicine*."
—*The American Prospect*

SICK

SICK

THE UNTOLD STORY OF AMERICA'S HEALTH CARE
CRISIS—AND THE PEOPLE WHO PAY THE PRICE

JONATHAN COHN

HARPER PERENNIAL

NEW YORK • LONDON • TORONTO • SYDNEY • NEW DELHI • AUCKLAND

HARPER ● PERENNIAL

FIRST HARPER PERENNIAL EDITION PUBLISHED 2008.

Designed by L. Pagnozzi

The Library of Congress has catalogued the hardcover edition as follows:
Cohn, Jonathan.
    Sick : the untold story of America's health care crisis—and the people who pay the price / Jonathan Cohn.—1st ed.
        xvi, 302 p. ; 24 cm.
    Includes bibliographical references (p. [233]–288) and index.
    ISBN: 978-0-06-058045-2
    ISBN-10: 0-06-058045-3
    1. Medical care—United States. 2. Medical care, Cost of—United States. 3. Health care reform—United States. 4. Poor—Medical care—United States. I. Title.
    RA395.A3 C635 2007
    362.10973 22   2006049236

ISBN 978-0-06-058046-9 (pbk.)

12 ID/RRD 10 9 8 7 6

FOR MY GRANDPARENTS:
ROSE, CHARLIE, BINKA, AND KURT

CONTENTS

INTRODUCTION

BOSTON

It was 4:43 on a clear November afternoon when the paramedics found Cynthia Kline, pale and short of breath, slumped against a bedpost in her double-decker Cambridge home. Although Kline was in obvious pain, she seemed keenly aware of what was happening inside her 55-year-old body. One of her blood vessels had closed off, blocking the flow of blood to her heart. Minutes before, she had phoned 911, taken the nitroglycerin tablets prescribed for such an emergency, then waited for help to arrive—an ordeal that stretched out an agonizing extra few seconds while the rescue workers, having found the front door locked, scampered in through an open second-story window. Now, while the paramedics worked busily over her, noting vital signs consistent with cardiac distress, Kline turned to one of them with an anxious plea: "Take me to Mount Auburn Hospital."

Kline, a teacher who worked with special-needs children, had no formal medical training. Yet her instinct about where to go was as sound as a seasoned cardiologist's. Nearby Mount Auburn Hospital, a private teaching facility affiliated with Harvard Medical School, had some of the city's finest doctors and nurses. More important, it had an intensive cardiac care unit that specialized in cases like hers. A few days earlier, staff at Mount Auburn had treated Kline's advanced coronary disease by inserting a balloon into her circulatory system and then expanding it, in order to open up a partially blocked blood vessel. A variant on the very same procedure, "cardiac catheterization," could be used in an emergency like this one, when the flow of blood through a vessel was almost completely cut off. Cardiac catheterization had saved literally thousands of lives across the country. The procedure had the potential to save Kline's life, too, just as soon as she could get to the hospital and receive it.

But getting there was precisely the problem. On the way to Kline's home, the ambulance driver had checked with a dispatcher about hospital availability. Mount Auburn was no-go: the emergency room there was overflowing, with no space to handle new patients. So as the paramedics wheeled Kline into the ambulance, one of them told her they would have to deny her request: "Ma'am, we're going to Cambridge Hospital instead."

Kline accepted the news, and maybe for a moment she thought it would be for the best. Although Mount Auburn was less than two miles away, Cambridge Hospital was even closer—just a short trip through the crooked, disjointed streets that surround Harvard Square. It was also a highly regarded medical facility in its own right, with a top-notch medical staff and a recently renovated emergency area fully capable of handling the majority of trauma cases that came its way. Had Kline's condition remained as it was, it probably could have handled her case, too.

But just four blocks into the journey, Kline's condition suddenly deteriorated. The instruments tethered to her arm could no longer detect a blood pressure; her heart rate, seventy beats per minute just moments before, was down to thirty-eight. Kline, strapped into a stretcher, was conscious through all of this—and increasingly agitated. At her side one of the paramedics, a kind-looking thirty-year-old, tried to calm her, explaining that the hospital was just seconds away. But as the ambulance made a right turn around one final corner, bringing the tall redbrick facade of Cambridge Hospital into view, Kline began to cry out: "I'm going to die. I'm going to die."

It was 5:04 p.m., just twenty-nine minutes after Kline had first called 911 and about an hour into the heart attack, when the green-and-white ambulance pulled up to the emergency bay. Informed of the patient's newly worsened state, attendants hustled the gurney into the hospital as the medical team began administering intravenous medication to increase Kline's heart rate. For a while it looked like she might pull through. Her pulse went back up to forty-five beats per minute—a far cry from normal but at least not "very low," as it had been in the back of the rig. Her breathing was more regular, too. Soon, however, a cardiology exam confirmed that Kline needed catheterization, something the staff at Cambridge Hospital could not do.

A nurse began inquiring about available hospitals, but now it was two hours since the chest pains had first begun—and time, finally, was running out. At 6 p.m. Kline's heart stopped altogether. The doctors began performing the familiar ritual of cardiopulmonary resuscitation (CPR), pumping her chest and using electrified paddles to shock the heart back into a regular rhythm. It made no difference.

At 7:03 p.m., the trauma team relented. Cynthia Kline was dead.

Fifty-five-year-old women, particularly those who have a family history of coronary disease, die from heart attacks all the time. So as a forensic matter, at least, Kline's death was unremarkable. But the technology that might have kept her alive existed—and it existed at a hospital that was less than five minutes away from her house. There was no guarantee that Mount Auburn's doctors could have saved Kline. Still, as one source familiar with the case told the *Boston Globe*, whose story on the matter sparked a state investigation, "Within an hour and a half they would have started to open her artery with a catheter. If you get the artery open there's a 50-50 chance."

None of which would be so troubling if the overcrowding at Mount Auburn on that day in November 2000 were an isolated incident. It wasn't. During a one-week period shortly after Kline's death, a survey of seventy-six Massachusetts hospitals found that sixty-seven of them had used emergency crowding procedures or had diverted ambulance traffic. Massachusetts General Hospital (MGH), Boston's largest medical facility, was closing its emergency room to patients forty-five hours per week. On the day of Kline's heart attack, MGH was the next closest hospital with a cardiac catheterization unit—just three miles away. But it wasn't taking new emergency patients that day, either.

And even if Kline was the city's only known fatality from ambulance diversion, there was plenty of reason to think that the overcrowding epidemic was routinely jeopardizing the well-being of patients. In 2000, when the Massachusetts College of Emergency Physicians surveyed the directors of more than sixty emergency room (ER) facilities, four out of five said they'd diverted traffic at some point—and nearly 40 percent said overcrowding had led to "adverse outcomes." Sometimes it was a matter of forcing ambulances to drive longer distances in order to find available hospital beds, or, as in Cynthia Kline's case, of shunting people to hospitals less able to provide advanced treatment. And sometimes it meant that patients who got to the right emergency rooms had to wait many hours before receiving treatment. In some cases, patients actually had to wait inside the ambulances.

One thing was certain, though. The crowding problem made little distinction among patients of varying status, wealth, or influence. Bob Maher was the chief executive officer of Worcester Medical Center in central Massachusetts when he had a heart attack in November 2000, during an airplane flight to Boston. Paramedics met him at the airport, but his connections weren't good enough to get him into MGH, which was, once again, on diversion. So the ambulance took him to another hospital several miles away. A woman named Nancy Ridley had her own troubles in the ER in May 2001. Suffering from a high fever and a hacking cough, she spent five hours waiting to be admitted for pneumonia at the Lahey Clinic in nearby Burlington. Ridley suffered no major health setbacks because of the wait, but the all too typical delay was the kind of problem she might have reported to the Massachusetts Department of Public Health—if only she hadn't already been working there, as its assistant commissioner.

Boston, in other words, had an emerging public health crisis on its hands. And it wasn't alone. In Atlanta, an ambulance crew carrying a patient in respiratory distress had to pull over and wait on the side of a highway for eighteen minutes because the nearest hospital was full and the paramedics were busy trying to find an alternative. Only when the patient went into full arrest—that is, he stopped breathing altogether—did the closest hospital find a way to take him. That patient lived, but others were not so fortunate. When the mother of a forty-year-old Cleveland man with liver failure called the local community hospital, staff there referred him to the MetroHealth Medical Center, which had more advanced lifesaving facilities. But when the ambulance arrived, MetroHealth was on diversion. The man ended up back at the community hospital, where, fifteen hours later, he died. In suburban Houston, when a twenty-one-year-old man was hit by a car, the local trauma centers turned him away because they had no room. He ended up on a helicopter ride to the next closest trauma hospital—in Austin, more than 150 miles away—and died shortly after arrival.

Fed up with incidents like that, a Texas neurosurgeon named Guy Clifton started an advocacy organization called Save Our ERs. The group's first order of business was to compile data on Houston's emergency services, and it produced sobering results. The city's two level-one trauma centers, the hospitals capable of dealing with the most dire emergencies, were simultaneously on diversion for extended periods once every two days. "I've been in this business for twenty-five years," Clifton said, "and I've never seen anything like this."

The United States has not had a serious political discussion about health care reform since the early 1990s. But if the situation in our emergency rooms is indicative, then perhaps it is time for another one. Overcrowding in ERs, according to most experts, is actually a symptom of other systemic problems now plaguing medical care—from the downsizing of less profitable hospital services such as psychiatric wards, where emergency rooms must frequently send patients who need admission, to the swelling ranks of people without health insurance, whose untreated chronic conditions are more likely to become serious medical crises.

To the casual observer, these trends might seem unrelated. But they are all consequences of the way Americans pay for their medical care—and of how that system is now falling apart.

It's a system of public and private insurance programs, supplemented by private charity, that dates back to the late 1920s—the time, not coincidentally, when medical care first became so expensive that large numbers of Americans literally could not afford to get sick. And it's a system that has survived for as long as it has because, by the late twentieth century, it had financed a massive industry dedicated to medical care while putting its services within reach of the majority of Americans. As critics have repeatedly noted, these arrangements have never met everybody's needs; the poor, in particular, have frequently struggled to find medical care either through doctors or through safety-net hospitals. But the U.S. health care system has generally worked well enough—or, more precisely, it has worked well for enough people—to withstand efforts at redesigning it.

Probably never was this more conspicuous than in the early 1990s, when President Bill Clinton proposed his now infamous reform plan. Under Clinton's proposal, the government would have made certain everybody had insurance coverage and, along the way, refashioned the whole health care industry—doing for Americans what the Canadian, Japanese, and western European governments have long done for their citizens. But Clinton's gambit failed. And while many critics would later blame its demise on either Clinton himself or the special interests that fought him, a more crucial impediment to reform may have been public ambivalence.

Most Americans, after all, still had health insurance in the early 1990s and rather liked it the way it was. When they needed medical care, they got it. To these people, the possibility of losing insurance and the consequences that might follow just didn't seem real enough to warrant such a sweeping overhaul—particularly if it would be at the hands of the government, an institution few people believed was capable of such a massive and complicated undertaking. "I've got pretty good health care and 80 percent of the country has pretty good health care," said one caller on a CNN show in August 1994, summing up a national mood that had turned decidedly against comprehensive reform. "Why are we doing the wholesale changes?"

And yet if Americans truly believed they had rejected radical change with Clinton's health care plan, they were in for a surprise.

The arrangements for financing medical care, from the private insurance workers got on the job to the public insurance programs that provided for retirees, were already faltering, because they could neither control nor keep up with the rising cost of medical care. The strain was building not just on emergency rooms, but also on charity clinics and public hospitals. Sooner or later, something was going to give.

In retrospect, then, the real issue in 1994 was not *whether* America's health care system should change but *how*. Who would be in charge? Who would benefit? Who would suffer?

What follows in these pages is the answer to those questions, told through the stories of a few ordinary Americans who came to learn them firsthand.

ONE

GILBERTSVILLE

New York's Leatherstocking Country sits at the northern foothills of the Catskill Mountains, a few hours' drive from Manhattan. The name is a reference to the leggings that colonial settlers wore during the 1700s, but it was James Fenimore Cooper who immortalized the region in the early 1800s, when he used it as the setting for his five books about the frontier hero Natty Bumppo, a collection that later became known as the *Leatherstocking Tales*. To see the area today is to glimpse a landscape remarkably like the one that first captured Cooper's imagination: a "succession of hills and dales" rolling through the countryside; "beautiful and thriving villages" nestled in the "narrow, rich, and cultivated" valleys, with only the occasional gas station and roadside pizza shack to pierce the "romantic and picturesque character." For the people who live in Leatherstocking Country now, this largely unmolested geography provides precious insulation from the rest of New York—even, it would seem, from modernity itself.

Perhaps nowhere is this more evident than in the village of Gilbertsville. Gilbertsville is a two-hour drive from Syracuse, the closest city
with a major airport, and the last part of the journey takes place along
a winding road, more than 1,000 feet up in the hills. Gilbertsville becomes visible only after the final turn, when the road descends into a
valley, depositing drivers near the entrance to Major's Inn—a Tudor-
style building that hosts such events as the annual Gilbertsville Quilting
Fair. Here Gilbertsville's commercial district begins and nearly ends,
with a small grocery store, a quilt shop, and a few offices operating
out of more Tudor-style storefronts that line one side of the street. The
buildings look almost precisely as they did when they were built in
the 1890s, after a fire destroyed the old business district. The effort to
replicate a small English town was apparently the inspiration of Joseph
T. Gilbert III, whose great-grandfather, Abijah Gilbert, helped settle the
township in the 1780s after migrating from Great Britain.

Some of Abijah's direct descendants still live in Gilbertsville. In
fact, say the locals, many of the 377 people that the 2000 U.S. Census
placed in Gilbertsville have roots in the area that go back at least 100
years. More than surnames were handed down over the generations.
The people in this part of New York have a long-standing reputation
for hard work, conservative values, and attachment to the land—a
reputation that still seems fitting today. The inhabitants attend church
regularly and strongly prefer conservative politicians, electing mostly
Republicans to the five-member village governing body. As for their
awareness of their heritage, perhaps the most celebrated episode in
Gilbertsville's modern past came in 1982, when its residents won a
seventy-year fight to block construction of a proposed dam that would
have flooded most of the village. They prevailed by methodically cata-
loging the architectural heritage of every local building, commercial
and residential, then successfully lobbying to have the entire village
placed on the National Register of Historic Places, forever protecting
it from disturbance.

It was four years after that victory that Gary and Betsy Rotzler moved to Gilbertsville, fitting into the community fabric almost seamlessly. They'd grown up together in neighboring Delaware County. Although Betsy was born in the Bronx, Gary's Leatherstocking lineage went back a dozen generations, to its very earliest days as a settlement in the New World. (Family legend had it that one of Gary's ancestors came from Britain to the United States on the ship immediately following the *Mayflower*.) The long hair Gary wore during his adolescence was typical for boys in the 1960s and early 1970s, but in most other respects he and Betsy were remarkably traditional. They had become high school sweethearts after going on a date to the county fair in 1975, while Gary was a junior and Betsy still a freshman, then continued dating after Gary went to college upstate. On June 25, 1978, just one day after Betsy's graduation from high school, the two were married in a small ceremony held at the home of Betsy's parents. A year later they would have their first child, a daughter named Sarah. Two more would follow: another daughter, Amanda; and then a son, Luke.

The Rotzlers came from relatively modest roots: Gary's father was a diesel mechanic who worked for the local highway department; Betsy's parents ran a residential treatment center for alcoholics. But by the time the couple came to Gilbertsville, the Rotzlers had every reason to expect they were on their way to realizing the American dream. Gary was noted for being industrious, having missed not a single day of classes in college. Shortly after graduating, he began working at Bendix, a large aerospace manufacturing company—first as a design technician at a plant in the nearby town of Sidney, later as a field engineer managing the company's midwestern clients from Dallas, Texas. Another engineering job at the Bendix Sidney plant had lured the Rotzlers back to central New York—where they hoped to remain, for good. Betsy, for her part, had chosen to stay at home and raise the three children, getting involved primarily in activities that revolved around them, like the La Leche League for mothers who were breast-feeding and, later, the Girl Scouts. Around town, she would become known for her artistic flair, particularly the individualized Raggedy Ann–style dolls that were her trademark.

But the American dream would prove fleeting for the Rotzlers, just as it did for much of central New York in the early 1990s. The regional economy depended on defense manufacturing jobs, like Gary's, that vanished as Washington cut the Pentagon budget and a recession fell over most of the country. One by one, Gary's colleagues lost their jobs. In 1993, he lost his, too. And while Gary would find ways to replace some of his lost wages over the ensuing two years, he would have a much tougher time coming up with something else: health insurance. Like most working Americans, Gary had always depended on his employer to provide medical coverage; when the job was gone, so was his coverage. And even after Gary finally found full-time work, he still couldn't get insurance for his family, because his employer—a company for which he'd worked previously—was no longer providing benefits to many of its employees.

None of this was unusual. On the contrary, it was emblematic of a change then taking place across the country: the erosion of job-based insurance, on which the U.S. health care system had been based for most of the twentieth century. The steady decline of job-based health coverage was the primary reason that the number of Americans without health insurance, nearly 40 million people by the time Gary and his family joined their ranks, was rising by the early 1990s. And, like Gary, the majority of these newly uninsured were neither destitute nor truly jobless. Instead, they were people who, as the saying goes, played by the rules of society—finding whatever employment they could, frequently working at several part-time jobs, but with no idea when they'd be able to get medical coverage again.

For some of these people, having no health benefits would eventually mean financial calamity. For others, it would mean a serious, even life-threatening medical crisis. For the Rotzler family of Gilbertsville, New York, it would come to mean both.

━━━━━

Although nobody knows for sure who invented insurance, historians generally trace its development back to ancient Babylonian traders who feared that their shipments across the desert might fall prey to bandits, dust storms, or camels with shoddy knees. In order to protect themselves financially, groups of these merchants decided that they would contribute to a fund; if a merchant's shipment disappeared, he could then take payment—or make a "claim"—to cover his losses. Later, the Greeks and Romans extended insurance beyond commerce by creating so-called benevolent societies, which pooled contributions from members to finance burials for the deceased. From there, the practice evolved into the mutual protection societies that the guilds of medieval Europe ran for their members, providing financial support in case of disabling injury or death. Eventually companies dedicated exclusively to providing insurance came into existence. The most famous among them was Lloyds of London, whose protection of merchants helped trade to flourish throughout the British Empire.

America's first private insurance company is believed to have appeared during the colonial era, when Benjamin Franklin established a firm to insure the homes of Philadelphia against the risk of fire. But it was not until the early twentieth century that the idea of using insurance to help people deal with illness started to get serious attention. At that time medicine was just entering what we now consider its modern era. With the development of sanitary techniques (to prevent infection) and sophisticated understanding of opiates (to dull pain), surgery had become more effective and widespread, turning hospitals from places where people were lucky to survive to places where people expected to be cured. Physicians, meanwhile, had developed formal

education and certification protocols, giving them a claim to expertise
beyond that of quacks and witch doctors. As one scientist of the era
famously remarked, "It was about the year 1910 or 1912 when it became possible to say of the United States that a random patient with
a random disease consulting a doctor chosen at random stood better
than a fifty-fifty chance of benefiting from the encounter."

But with this progress in medicine came new costs. Doctors and
hospitals expected to be paid well for their services, particularly since
they were investing so heavily in their training and equipment. And by
the 1920s, the bills were becoming more than many Americans could
bear. With the onset of the Great Depression, the average cost of a
week in the hospital began to exceed what the majority of Americans
earned in a month, making illness a scary financial proposition for
even the thriftiest middle-class households—and forcing many people
to skip medical care altogether. "Very few of these families are indigent in the accepted meaning of that word," the economist Louis Reed
explained in 1933. "They have a home, they buy their own food and
clothing and pay their doctor's bills in ordinary illness. But when a
serious illness . . . occurs, these families are unable to pay their way."

Reed had gotten his information from the Committee on the Costs
of Medical Care, a blue-ribbon commission that had spent five years
conducting the first national census on health care. But not all of its
findings were so bleak. In particular, the committee observed that
large expenses were concentrated among a small group of people, the
ones with the most serious medical problems. Since everybody had at
least some risk of experiencing such a crisis at some point in his or her
life, the committee recommended that Americans do what the ancient
merchants had done, and assume some form of collective responsibility for medical costs. In other words, it recommended the creation of
insurance for medical care.

The key question, of course, was how. Other industrial countries
were starting to give insurance to every citizen, through the govern-

ment or government-sponsored organizations, thereby spreading the financial burden of medical spending as widely as possible. Health care in these countries was thus on the way to becoming a right, rather than a privilege. But calls to do the same thing in the United States had run into stiff political resistance ever since the late progressive era, when state-level reformers in California and New York first proposed it. Large corporations feared that government management of medicine might lead to interference elsewhere in the private economy. Private insurers weren't ready to concede a possible line of business. And physicians simply didn't want the government meddling with their work or incomes. Physicians would prove a particularly potent lobby during the first half of the twentieth century, to the point where Franklin Roosevelt is said to have dropped health insurance from the Social Security Act because he feared that the hostility of state medical societies and the American Medical Association would undermine the whole initiative.

The physicians were so worried about outside interference, private or public, that many would have been content to do absolutely nothing about the financing of medical care during the 1930s. But it turned out that consumers weren't the only ones struggling financially. Hospitals needed help, too. In the years leading up to the Depression, while the economy was booming, hospitals had gone on a building binge, constructing new wings and outfitting them with the latest, most expensive equipment. Now all those brand-new facilities were either empty or full of patients too poor to pay their bills, presenting the hospitals with a crisis of their own.

One of those institutions was Baylor Hospital in Dallas, Texas, which by 1929 was "just 30 days ahead of the sheriff" because of its mounting debts. But Baylor also had a new administrator, Justin Kimball, with an idea for saving the hospital financially. Kimball, who had come to Baylor from the Dallas public schools, decided to approach his old colleagues with an offer: the hospital would provide up to twenty days of care to any teacher willing to pay a monthly contribution of fifty cents, so long as at least three-quarters of the system's teachers agreed to be part of the plan. Meeting that three-quarters threshold was crucial: Kimball, who had access to the school system's personnel records, had calculated that it would guarantee enough contributions from healthy teachers to cover the medical expenses of those few who needed care. But Kimball had no problem recruiting so many enrollees, as the promise of economic security for such a modest price turned out to be an easy sell. On the first day of Christmas vacation in 1929, a teacher who had broken her ankle on the ice showed up at Baylor's emergency room, becoming the first beneficiary to make a claim under modern hospital insurance.

As word of the success at Baylor spread, hospital administrators around the country copied the model and improved on it—eventually establishing nominally independent plans that paid for services based on fees the hospitals set. In Sacramento and later central New Jersey, hospitals joined together to create a plan that offered beneficiaries care at any local facility, not just one, thereby starting the nation's first multihospital insurance plan. In 1934, the founder of a plan based in Saint Paul, Minnesota, decided to illustrate his advertising posters with a blue cross. The image caught on as fast as the plans themselves, and by 1938, 2.8 million people were enrolled in "Blue Cross" plans that had established themselves across the country.

Like the original scheme at Baylor, most of the early Blues plans concentrated on offering coverage to groups of employees—or, occasionally, through fraternal societies like the Elks Club—because that was the best way to guarantee a group of subscribers sufficiently large to make the insurance math work. As commercial insurers got into the health business, they, too, focused on large workplaces. A habit was forming—one the government would soon make very hard to break. During World War II, federal officials decreed that fringe benefits were exempt from wartime controls on wages. That encouraged employers to offer more generous health insurance, since better benefits were one of the few enticements they could use to attract new workers in such a tight labor market (with much of the able-bodied workforce busy fighting overseas). Soon the government also decided that money spent on health insurance provided by employers would not be subject to the income tax. This increased the demand for such benefits—since, to a worker, a dollar of health insurance became more valuable than a dollar of salary.

Linking insurance to employment meant that businesses were, in effect, becoming responsible for their employees' well-being. But that was not a burden corporate America seemed to mind. On the contrary, businesses eagerly pursued this role because they believed it would cement workers' loyalty while undermining the appeal of national health insurance—something they continued to oppose on ideological grounds. Once the National Labor Relations Board ruled that health benefits could be part of collective bargaining agreements, labor embraced this arrangement, too—recognizing that it made the unions even more important to their workers. Labor didn't exactly give up on national health insurance; in the late 1940s, when Harry Truman formally proposed and campaigned hard for a universal coverage plan, the AFL-CIO supported him. But some unions were less enthusiastic than others. The United Mine Workers largely sat out the fight, in part because its leaders were pleased with the coverage they'd won for their members and thought that national health insurance, if anything, would lead to less generous benefits.

For another twenty years or so, that strategy paid off for the unions—and, indeed, for most Americans—as the percentage of the workforce receiving insurance through their jobs climbed. By 1980, most full-time workers at large companies had health insurance through their jobs. And the benefits kept getting more comprehensive, too. This encouraged people to use more medical services, which led to more health care spending overall. It also fueled additional investment in medical technology, which drove up the cost even more. But particularly since people seemed to benefit from all of the extra medical attention, nobody seemed to mind that much. The economy was growing so fast that businesses were willing to absorb the higher expenses on behalf of their workers (and the workers, in turn, hardly noticed the way health insurance premiums were eating into their paychecks).

But that neat arrangement started to break down, as the price of health insurance premiums skyrocketed and the long postwar prosperity came to an end. During the 1980s, large manufacturing companies, once the linchpin of the U.S. economy, were suddenly desperate to cut costs because they were struggling to keep up with foreign competitors who could produce goods more cheaply. Retail and service industries were faring better. But, in part because very few of them were unionized, they were less inclined to offer coverage in the first place.

As workers faced higher premiums, some opted against buying coverage even when it was made available to them. Others never even had that choice. Just as cheap health insurance premiums and tight labor markets once created pressure on all employers to provide their workers with decent coverage, now expensive premiums and loose labor markets created pressure not to provide coverage. By the 1990s, the success stories of the U.S. economy were primarily those corporations, like Wal-Mart, that squeezed employee benefit costs by offering skimpy benefits and limiting them to full-time workers who'd been employed at the company for at least two years. As of 2005 less than half of Wal-Mart's employees had company-sponsored insurance, well below the national average for large firms.

Those companies unable to scale back their commitment to finance generous health benefits, meanwhile, paid a steep price: In 1993,

General Motors said that health insurance for its employees alone added more than $700 to the price of every car and truck, or roughly the cost of putting air conditioning into a Chevrolet. By 2004, GM said that the figure was up to $1,400. And the company, not surprisingly, was rumored to be flirting with bankruptcy.

Gary Rotzler was among the individuals caught in this economic undertow. In the early 1980s, at the height of that era's mergers and acquisitions frenzy, Gary's employer—Bendix—had attempted to take over a rival aerospace manufacturer, Martin-Marietta. The episode turned into "one of the dirtiest, sloppiest, most wasteful takeover battles in U.S. corporate history," as one *Time* magazine writer put it. William Agee, the chief executive officer, walked away from the debacle $4.1 million richer; Bendix was left in the hands of another company, Allied Corporation, and never fully recovered.

Previously, Bendix had a reputation as one of central New York's best employers, known for its generous pay and its loyalty to workers. The employees, in turn, treated it as part of the community. Once they got jobs at Bendix, many figured they'd stay until retirement. But in the late 1980s, Bendix, then renamed Amphenol, began the inevitable downsizing that would eventually hit virtually every manufacturer in the state. Gary figured he could either get out or be tossed out, so in 1989 he left Amphenol to take a job at ShopVac, a company that had a plant in another nearby town, Norwich. The move actually constituted a promotion of sorts, to the job of quality engineer. Although Gary didn't know much about vacuum cleaners per se, the engineering work was actually quite similar to his work at Amphenol, since in both cases he was working on the production of plastic parts.

Rather than escape the day of reckoning, however, Gary had merely postponed it. One day in 1993, a new manager called Gary into his office and told him that he no longer had a job. And this time around, engineering jobs were not so easy to find. Gary would scour the newspaper for job openings and apply for every one, but in each case there seemed to be about 200 applicants against whom he was competing. He thought about applying for public assistance and even drove to Cooperstown, the county seat, so he could fill out the paperwork. But Gary says he couldn't bring himself to complete the forms—"I didn't want a handout; I wanted a job"—and he decided to keep searching.

Gary was as comfortable working with a two-by-four as he was at a drafting table, and within a few weeks he had found temporary work in a neighbor's construction business. For the better part of two years, he'd take on jobs remodeling houses and crafting furniture, attending to each with his familiar determination. (One summer he worked so hard that his weight came down from 220 pounds to 160.) At this time, Betsy rejoined the formal workforce, running a small day-care center at her home. Occasionally Gary and Betsy talked of putting their crafting skills together and opening a small business, with Gary building furniture and Betsy decorating it. But mostly Gary just kept interviewing for new jobs in electronics and defense firms, figuring he was bound to find one sooner or later.

At the end of 1994, he did—back at the Amphenol Sidney plant, doing work similar to what he had done in his previous stint there. But now he was designated a temporary worker, which meant no health insurance. Gary had no reason to believe the job was temporary in any meaningful sense: his primary job involved checking for flaws in electronic devices produced for the International Space Station, and Amphenol's contract with the space program meant it would be churning out the products for many years. Nor did it escape Gary's notice that many other "temporary" employees were people who, like him, had once held permanent positions at Amphenol.

But, according to Gary, his repeated requests to be reclassified as permanent and full-time, so that he could get health insurance for himself and his family, were unsuccessful. A mergers-and-acquisitions firm had acquired Amphenol from Allied, and managers frequently mentioned that they had strict department-by-department quotas for permanent employees entitled to benefits. Gary says he was told that because he was an "indirect employee"—which meant that he was responsible for monitoring the quality of products rather than actually making them—there was simply no way to add him to the head count.

By early 1996, Gary was putting in between sixty and seventy hours a week. In order to get home for dinner every night, he'd leave his house shortly after 5:30 a.m., getting to work at around 6 a.m. He'd also put in half days on weekends. It was a good way to make extra cash, while possibly impressing somebody in a position to give him a permanent position. But with all the time away from home, Gary would later realize, he was slow to detect a development there that his friends and families were noticing back home: something was wrong with Betsy.

———

The change was particularly evident to Betsy's younger sister, Trisha. One of the things Trisha remembered most vividly about her childhood was Betsy's sheer raw strength. Even as a child in elementary school, Betsy could literally lift most adults off the floor. She was always tall for her age, too. To Trisha, Betsy was sometimes more a big brother than a big sister, like the time she beat up a boy who had been teasing Trisha on the school bus. (Trisha says she didn't always welcome the protection; years later, at least one potential boyfriend lost interest once he found out that Betsy was her older sister. He told Trisha he was afraid Betsy might beat him up, too.) Betsy also had a loud, infectious laugh—more a belly laugh than a cackle, according to her sister—plus a sharp wit to go with it. Once Trisha had tried to sew a stuffed horse for her own daughter. But Trisha wasn't the artist that Betsy was. And when Betsy saw the somewhat disfigured creation, she laughed and told Trisha she should try selling it, with a label that said "Crafted in the Catskills by the blind."

Betsy was usually full of energy, too: "She was always in such a hurry to do things," Trisha would later explain. "Anything she wanted to do she wanted to go ahead and do right away." Trisha was hardly surprised when Betsy got married right after graduating from high school, just as she was hardly surprised when Betsy had her first baby only a year later. And although Betsy had forsaken a career for motherhood, she frequently talked about returning to school once the kids were older and then finding outside work again—maybe as a developmental psychologist. Caring for half a dozen kids in her house intensified that interest. Betsy would frequently call Trisha to share little stories about something the kids had done that day.

By 1996, however, Betsy sounded beleaguered every time the sisters spoke. She didn't talk about her future or her plans for decorating the house, which were now on indefinite hold; when the discussion turned to the Rotzlers' kids, she worried that they didn't have enough new clothes or that the house was always cold because she and Gary were trying to cut their heating bills. Betsy's younger brother, Steve, picked up on the change, also. For the first time that he could remember, Betsy was complaining about the winter, saying over and over again that she and the family had to "get out somehow."

At first, Betsy's family assumed she was just reacting to the financial stress and Gary's long hours away from home. But then Betsy, who had been complaining of flu-like symptoms for several weeks, started talking about something more specific: a pain in her back. First she described it as a dull pain, like a chronic backache. But it got progressively more intense every month. One day in the summer, while the whole family was together, Betsy's mother, Mary Lou Harvey, tried to give her a back massage to ease the discomfort. Mary Lou—who had a slight, fragile build—had barely touched her daughter's spine when Betsy screamed, "Ouch!"

———

As extroverted as Betsy was with her loved ones, she was frequently shy with outsiders, including the townsfolk of Gilbertsville. She preferred doting on her children at home to gabbing with her neighbors at the grocery store—partly, her neighbors gathered, because she could be self-conscious about her appearance. For all her youthful displays of boyish brawn, Betsy was still, by most accounts, a "girl's girl" who liked to put on makeup and shop for clothing. With deep blue eyes and full red hair, she was striking when her five-foot-ten frame was down to 120 pounds, as it was when she married Gary. But her weight had fluctuated a great deal once she started having children; sometimes she would put on twenty or thirty pounds. She could usually shed them when she put her mind to it. Compliments from neighbors, though, frequently prompted Betsy to sink back into her clothing, as if she were embarrassed that anybody had noticed the gain in the first place.

The one side of her that locals did see was her generosity—a generosity that had continued even when she and Gary didn't have much for themselves. Jean Postighone of nearby Morris first got to know Betsy because they kept bumping into each other at the same craft store, whenever Betsy came by to pick up supplies for her sewing. They eventually began working together, until one day, after Jean was bemoaning the fact that she'd never found the time or money to decorate her windows, Betsy offered to do it for her. With some extra fabric picked up at J.C. Penney, Betsy sewed new window treatments for the entire first floor, accepting not a dime for her services. That was typical of

Betsy, neighbors agreed; she probably could have made good money selling her hand-sewn dolls if only she hadn't kept giving them away.

Jean was among the neighbors who, sometime in early 1996, began to see that Betsy's appearance had worsened. Her weight was back up and her usually youthful face had a weary, aged look. When Jean would stop by the Rotzlers' house, Betsy—typically so attentive to her looks—would appear in a housecoat, with no makeup. Other neighbors also noticed changes. Betsy seemed unusually tired, and she was even more reluctant to venture outside than before.

Betsy's neighbors were hesitant to pry, since she obviously valued her privacy highly. Trisha, on the other hand, had no such inhibitions. Certain that something besides stress or fatigue was ailing her sister, Trisha over the spring and summer urged Betsy to seek medical help. But Betsy had already made one effort, calling the office of the obstetrician who'd delivered two of her children; she had apparently been told that, since she had no health insurance, she would have to pay the full price for the obstetrician's medical services up front. With finances tight, Betsy thought she and Gary needed to save what money they had for real medical emergencies—such as Amanda's leg injury in 1995, which ended up costing $600. "Every time the children got sick, they were very diligent about that—they would pay for it," said Mary Lou. "But every visit to the doctor was $150, average. They were managing, but every time somebody got sick it blasted them out of the water again."

So instead of getting professional medical help, Betsy tried alternative remedies. She dabbled in herbal medicine and even saw a healer. She also started sleeping on the couch, because it was easier on her back than the bed. She hadn't given up on seeing a doctor altogether, she assured Trisha. But with Gary interviewing for jobs every week, she figured she could wait until he finally found one.

When Americans boast that they have the world's greatest health system, they frequently have in mind what they've seen on television: the mind-boggling new treatments that American medical schools and hospitals produce. And few experts would dispute that the U.S. health care system has been enormously successful at pushing the boundaries of medical technology. Magnetic resonance imaging (MRI), now an essential diagnostic tool in virtually every major medical specialty, was developed in the United States. So was the equally valuable computed axial tomography (CAT) scan, not to mention a host of chemotherapy protocols.

But neither high-tech scanners nor more pedestrian forms of medical care do much good to people who lack access to them. And while most Americans assume that people without health insurance still get good health care, a mountain of data suggests otherwise. The uninsured are less likely to have regular doctors, and they see doctors less regularly. As a result, they miss out on a lot of the routine screenings that insured Americans get, such as mammograms, colorectal screenings, and simple cholesterol tests. Not all these procedures are necessarily as worthwhile as they may seem; recent studies, for example, have suggested that widespread screening for prostate cancer has turned up a lot of borderline, slowly spreading growths for which men were persuaded to receive invasive, sometimes debilitating treatment that wasn't really necessary. By and large, though, the screenings and tests that the uninsured don't get are precisely the ones that might help them avoid more serious health problems in the future— problems that incur serious financial and physical costs.

Betsy was, in this sense, entirely typical: By the fall of 1996, she had gone more than two years without a regular medical exam. And because of her fatigue and back pain, she had learned to live with what she assumed were less severe medical problems—among them, a chronic condition called *hidradenitis suppurativa*. The condition causes carbuncles (large, pus-filled lumps) to appear around glands;

women who suffer from it frequently get these carbuncles underneath their breasts, usually right around when they are menstruating. The carbuncles can be painful, but with rare exception they are harmless. And while doctors frequently treat them with antibiotics, they also will go away on their own. That is what Betsy apparently assumed would happen when, one day in September, one of the reddish lumps appeared on the underside of her breast.

But this lump, unlike the other ones, didn't go away. Instead, it got more uncomfortable. Finally she told Gary about it—and leveled with him about the severity of the pain she'd been feeling for months, something she'd been playing down. Realizing for the first time that Betsy might have a serious medical problem, Gary hurriedly made phone calls and consulted friends, one of whom told him that free exams were available at the Oneonta Planned Parenthood clinic, a controversial family planning center and abortion provider. During a hastily arranged appointment, the doctor examined Betsy for only a few minutes before taking her aside, privately, to deliver the news: she might have breast cancer.

Now things began to happen quickly. Gary arranged to have Betsy admitted to the Fox Hospital, also in Oneonta, for a biopsy and MRI. Betsy stayed overnight and Gary, still shell-shocked, figured that this was where she'd ultimately get treatment, too. But over the next week, as Betsy started telling people what was happening, family members told Gary he needed to take her to one of the big hospitals in Manhattan. Trisha, who was a writer, took it on herself to find a specialist on the Internet, eventually tracking down an oncologist at the Memorial Sloan-Kettering Cancer Center who was running a clinical trial that seemed designed for people in Betsy's condition. After the oncologist agreed to see Betsy, she and Gary decided it was best to leave Sarah, Amanda, and Luke—who by then were seventeen, thirteen, and nine—back in Gilbertsville. Among other things, Sarah was supposed to direct a high school play that weekend, and they didn't want her to miss it. Friends agreed to take them in.

A few days later, as the couple made the four-hour drive downstate in Gary's aging Chevy S-10 Blazer, Betsy had to lie down on the rear bench for the entire ride because her back hurt so much. Both Rotzlers feared that Betsy's condition was grave, although they didn't realize exactly how grave until they finally got to the oncologist's office. On the phone, the doctor had sounded slightly hopeful; even if Betsy's cancer was terminal, treatment might at least extend her life by a few years. But after he took just one look at the MRI, the prognosis began to turn grim. The cancer had metastasized to Betsy's spine; that was why she had been feeling such intense pain. It had also spread to her liver.

Betsy was admitted to Sloan-Kettering's hospital immediately, making the one-block journey from the physician's office in an ambulance because she was too weak to walk. She wouldn't be participating in the clinical trial, because the cancer was too advanced. Instead, Betsy's doctors began a regimen of intensive chemotherapy, delivered directly to her circulatory system through a stent. Combined with morphine to dull the pain, the treatment knocked Betsy into semiconsciousness. Betsy's siblings and parents had descended on the hospital, by now, and Trisha did her best to direct Betsy's attention, asking her to concentrate on her own daughter, Betsy's niece, who was also at the bedside. Betsy tried to focus her eyes but couldn't. With the strength drained from her body, she lapsed into a coma.

The next day, a Saturday, Gary dispatched Betsy's father to bring the children from Gilbertsville. Since arriving at Sloan-Kettering three days before, Gary had maintained a vigil at Betsy's bedside, except for one trip a day for food, usually at night when the cafeteria and nearby restaurants were least crowded. This night, after a quick trip to McDonald's, he noticed that Betsy's breathing had changed, with long pauses between breaths. He had to step away from the room momentarily, in order to sign some consent forms, including one promising to pay Betsy's medical bills if no insurance would cover them. He scribbled his name and returned to notice another change: in his absence, hospital staff had moved the room's other occupant elsewhere. The respiration sounds he'd detected, doctors would later tell him, were the sounds of Betsy entering her preterminal breathing pattern. This went on for a few minutes or maybe longer—Gary really isn't sure anymore—until the breathing stopped altogether. Betsy was dead.

Gary collapsed into a chair. Betsy's relatives wept. Even the doctor who had been taking care of Betsy in the hospital, a young Russian intern, seemed devastated by her rapid demise. On Sunday morning, the Rotzler children arrived at a relative's house just outside the city, where the family had gathered—still unaware of what had happened, because Gary wanted to break the news himself. Betsy's mother and siblings watched from the window as the three kids piled out of the car, clutching colorful "Get Well" balloons for their mother. Moments later, they learned that she was gone.

———

Gary didn't want to contemplate finances at this point, but he really had no choice, since now there was a funeral to pay for. Gary figured that his net worth couldn't be much more than the house, the $300 in his pocket, and the ten-year-old SUV in the parking lot. Betsy had no life insurance policy to cover the costs of interment; nor had she been in the workforce long enough for Gary to collect survivor's benefits from Social Security. Gary ended up borrowing the funeral money, putting up part of the house as collateral, and even then the funeral was a pretty minimalist affair. The undertakers hadn't applied much makeup to Betsy's face, and, upon spotting the open casket, a friend exclaimed loudly, "Oh, my God, she looks like Mrs. Doubtfire." That raised a few eyebrows among the well-wishers, although Trisha just laughed. It was, she figured, exactly the kind of joke Betsy would have made—right before erupting into one of her hearty laughs.

Another problem for Gary was that he still didn't have health insurance. When he first learned that Betsy might have cancer, he had stopped asking Amphenol for a permanent job and had started begging: "I had offered to push a broom when Betsy got sick. I offered to do anything." But Amphenol couldn't find a place for him—until one month after Betsy died, when a position opened up in sales. By then, the bills from Betsy's care were already piling up. Gary owed $40,000, most of it to Sloan-Kettering. He had no way to pay.

As word of Betsy's death and Gary's financial crisis spread around Gilbertsville, the village rallied. Neighbors threw a spaghetti dinner fund-raiser. Men from a local heating company came by to fix the furnace without charge. The local YMCA offered to provide free day care for Luke. But it wasn't enough. A year and several dozen collection calls later, Gary decided to declare bankruptcy. He got to keep his house, but his credit rating was ruined. He couldn't keep so much as a credit card, to say nothing of cosigning a loan so that his children could go to school.

Such financial devastation was not unusual by the late 1990s. On the contrary, it was one of the telltale signs that private health insurance in the United States was falling apart. According to research by Professor Elizabeth Warren of Harvard Law School, medical debt had become a leading cause of bankruptcy in America. A large fraction of the bankruptcies came from Americans without health insurance who ran up five-figure or even six-figure hospital bills, sometimes during just one emergency or episode requiring intensive hospital care.

Years later, people close to Betsy couldn't help wondering whether she had been in denial, particularly during those last months when her condition was clearly worsening—unwilling to believe she had a potentially life-threatening illness or, perhaps, unwilling to contemplate the possibility of what it might do to her family. "Obviously, I'm sure she was just hoping it would go away; she was hoping nothing would happen" Gary would say later. "If she got sick she knew what would happen: it would bankrupt us." Trisha suspected that Betsy was probably too embarrassed to say anything when her obstetrician's office demanded payment, even though the obstetrician was a family friend. "If the doctor himself had known, he would have seen her. But she probably talked to the office staff." It was all such a shame, one close friend added, because "when you go to the hospital, they will still treat you; it doesn't matter whether you have a million dollars or five dollars—if you are sick, they have an obligation to treat you."

The friend was correct, in one sense: doctors and hospitals will treat anybody with an emergency, regardless of ability to pay, if only because a 1986 federal law requires them to do so. And partly because that fact is relatively well-known—hospitals have to post signs about it prominently in waiting areas—many Americans believe that the uninsured can always find care if they really need it. As the Institute of Medicine, an independent, government-charted research organization, noted in a 2002 report about the uninsured, "In this country, we do not see many people dying in the streets because of inaccessible health care, and it has been easy to assume that people without health insurance manage to get the care they need."

The problem with that assumption, though, is what happens before patients get to the emergency room. And breast cancer, which took Betsy's life, turns out to be one of the most vivid examples. A patient diagnosed with stage-zero breast cancer has a 95 percent chance of living for at least ten years; for a patient diagnosed with stage-one breast cancer the chance is 80 percent. From there the chance declines steadily. If a patient has metastatic breast cancer—a cancer that has spread from the breast past the lymph nodes to the bone, liver, or lungs—she is virtually certain to die. According to one study originally published in the *New England Journal of Medicine*, women without health insurance were more likely than woman with insurance to be diagnosed with later-stage tumors—and, as a result, were nearly 50 percent more likely to die.

What's true of breast cancer is true of other cancers and, more generally, other diseases. Uninsured Americans with a chronic illness, whether it be diabetes or hypertension, are less likely to get the regular checkups and treatments their condition requires—and end up in worse health.

To be sure, nobody can ever know for certain whether regular medical care could have pushed Betsy's cancer into remission, or at least contained it long enough to give her a few extra years of life. But it would certainly have improved her chances. If the uninsured are not dying on the streets, that's probably because they are dying in hospitals—where they never should have been in the first place.

———

Closure for Gary came slowly. He kept the salt-and-pepper beard he'd grown in the hospital and gained back some of the weight he'd lost, so that people started commenting on his striking resemblance to Michael McDonald, the singing star who first gained fame as lead vocalist for the Doobie Brothers. Gary also started venturing out into Gilbertsville a little more frequently—he became "a walker," in the local parlance—although he remained a bit introverted. During the first year after Betsy's death he drew modest support from weekly meetings with a church pastor. Later he was able to work through the familiar stages of grieving via an online community for widowers, in part by posting a series of poems. Some were angry, lashing out at "death you foul beast"; others expressed survivor's guilt, asking "But, Why not me Lord, why not me?"

Betsy's mother, Mary Lou, channeled her feelings a little differently—by joining a local political organization and becoming an activist on behalf of health care consumers. "All I wanted at that point was to try helping people avoid the same experiences we had," she explained. Gary, although uninterested in becoming a spokesman, nevertheless came to see things in much the same way. "When you walk into a doctor's office and you've got to say you don't have insurance, there is a mood swing, there is a change, no matter how nice that person is behind the desk. You're a different class of person."

Gary, who had stopped voting Republican in the 1990s, made those remarks ruefully in late November 2004, just after the presidential election. Before Election Day, it seemed as if the issue of the uninsured—on political hiatus since the Clinton health care fight—was finally coming back. The ranks of the uninsured had, briefly, stopped growing in the late 1990s, thanks in part to the strong economy. But by 2001, when another recession hit, many people began losing coverage again.

The political system took notice, and in 2004 Democrats seemed to be shaking off the political hangover of "Clintoncare." Every major presidential candidate in the Democratic Party proposed ambitious initiatives for health coverage; although none proposed truly universal coverage, as Clinton had, several proposed schemes that would, theoretically, have left just 10 million people without coverage—a dramatic reduction.

But the Democrats didn't win in November. George W. Bush did. His proposals to help make health care more affordable consisted of tax credits for the purchase of insurance, which experts almost universally agreed were too small to have much impact; reform of medical malpractice laws, which scholars had repeatedly and convincingly rejected as a major cause of declining health insurance coverage; and support for replacing comprehensive coverage with catastrophic-only insurance, which threatened to destabilize further the job-based insurance system. Even the most optimistic—and, by most accounts, unrealistic—assessments of the Bush administration's own economists suggested that these efforts would make health care more accessible for just a few million people, or just a small fraction of the uninsured population.

But the merits of Bush's ideas were perhaps less significant than the political emphasis he gave them. Although he had sometimes mentioned them in campaign speeches and would go on to include them in two successive versions of the State of the Union address, he never made expanding health insurance a priority—preferring to commit his domestic policy money to tax breaks that favored the very wealthy.

By both word and deed, it was apparent that Bush, like the kindred ideological spirits who prevailed during the fight over "Clintoncare," simply didn't think the health care crisis was real. Millions of Americans knew otherwise. But, just like the Rotzlers, they were left to face the crisis on their own.

TWO

DELTONA

For nearly two years, Janice Ramsey had been looking for health insurance. And for nearly two years, she had been failing to find it. Her problem wasn't lack of a job. She had one of those. And it wasn't lack of money. She had some of that, too. No, Janice's problem was lack of health. She had diabetes.

It was a problem because Janice happened to be self-employed: she was a consultant to a home construction company she had once run with her husband. The work paid well enough. The company was based in Deltona, a booming suburb north of Orlando; and there was, if anything, more work than she could handle. But because she was working on her own, she also had to buy insurance on her own, rather than through a large employer. As Janice discovered, insurers won't usually sell coverage to individuals who have preexisting medical conditions that generate high medical bills. And few conditions spook insurers more than diabetes.

Janice says that many insurers, once they heard she was diabetic, told her not to apply at all. The few that considered her application either declined it outright or refused to cover anything related to her diabetes—rendering the policies pretty worthless, Janice figured, since so many medical problems could plausibly be blamed on the condition. "If I would have had a heart attack or anything else, if my foot fell off," she explained in her characteristically tart way, "they would have told me it's from diabetes because everything has to do with diabetes from what the insurance companies feel."

As the months dragged on and Janice's medical bills slowly depleted her savings, she frequently wondered how she had ended up in this situation. At the time, Janice was in her fifties. Having worked for much of her adult life, she carried herself like a professional, dressing in elegant business clothes and carefully styling her lightly colored short hair. Along the way, she had also managed to raise five children, all of them now in college or in successful jobs. Just recently, Janice had taken in her own mother, who had become too sick to live alone. Janice had a hard time imagining that many people worked harder than she did. And yet here she was, no better than a pauper as far as American health care was concerned. "It's embarrassing," she later said, "because I've never asked anybody for anything and I don't like not being able to take care of myself."

That's why Janice was so pleased when, in the summer of 2001, a friend told her about a new health insurance company called American Benefit Plans. American Benefit operated through professional associations (like realtors and photographers) to bring the advantages of large-group coverage to people working on their own or in small business. And on the basis of the plan's glossy literature, it sounded like a great deal. American Benefit had its own network of well-respected doctors and hospitals, including some of the best in the Orlando area. The monthly premiums were reasonable—a lot lower, in fact, than Janice had come to expect. Best of all, American Benefit didn't subject applicants to individual medical assessments or exclude coverage of preexisting conditions. In other words, it would cover her diabetes.

Janice signed up, and almost immediately the company began deducting the monthly premiums of around $365 directly from her bank account, as she'd authorized it to do. For a few months, the coverage seemed to be everything she had hoped it would be. She was able to take care of her routine needs, from the medication she used to help control her blood sugar to the checkups diabetics are supposed to get four times a year. She was able to take care of more serious problems, too, including a hospitalization after she collapsed in her home. Fearing that Janice was having a heart attack, the doctors had gone ahead and performed cardiac catheterization. It turned out they were wrong; Janice was instead suffering from severe exhaustion, most likely from the combination of work and caring for her mother. Still, she was pleased to have gotten such attentive medical care—particularly since, as with all her other medical needs, the insurance had covered it completely.

Or, at least, that's what the insurance was supposed to do. Seven months later, as she remembers it, a letter arrived announcing several thousand dollars in unpaid charges. Janice called the hospital; when the staff there told her the bill hadn't been paid, she told them to resubmit it. "I have hospitalization coverage," she told them. "This must be a mistake." When the hospital contacted her again, explaining that it still hadn't been paid, Janice decided to call American Benefit. There, a clerk told her that the company was simply reviewing the claim before paying it. That sounded reasonable enough—insurers did that all the time, she knew—so she didn't press the matter.

But American Benefit didn't pay. And when the hospital stopped sending bills and started dispatching collection agents, Janice decided she needed to bring in the authorities. She called the Florida Department of Insurance, hoping she could persuade it to compel American Benefit to pay up. And that's when a state official broke the bad news to her. American Benefit was not even licensed to sell health insurance in the state of Florida. The operation was a scam.

Janice would later learn that she had plenty of company. In a period of roughly two years, thousands of people across the country had bought phony coverage from con artists running similar operations. State officials eventually succeeded in shutting down these scams. They even put some of the perpetrators in jail. But they were less successful at recovering the victims' premiums—money the victims desperately needed because they now owed, collectively, millions in unpaid bills to doctors, hospitals, and other health care providers.

It was not the first time this had happened. On the contrary, this was the third wave of similarly designed health insurance scams to hit the United States since the 1980s. And although the victims included a wide variety of Americans—urban and rural, affluent and poor—they all shared one significant characteristic: they worked either for themselves or for small businesses and, as a result, had trouble finding affordable health insurance in the legitimate market.

It was a problem inherent in the nature of private insurance, which had evolved around the needs of large employers. But it was also a problem that had gotten conspicuously worse in the last twenty-five to thirty years, leaving scam victims like Janice Ramsey in a bind: up to their ears in medical debt and without the insurance to pay new bills. People like Janice had enrolled with American Benefit because they believed it would save them from mounting medical bills. Instead, it made their problems worse.

How does a country that supposedly lionizes the entrepreneur make it so difficult for him—or, as in Janice Ramsey's case, her—to buy health coverage? One part of the answer is sheer scale. Imagine the difference, for an insurance company, of trying to deal with 500 independent customers rather than a business with 500 employees. To work with those 500 independent customers, the insurer must first sell each person a policy through an agent, who will in each case demand a commission. The insurance company will then have to handle each subscriber's payments and claims separately—tasks that will require hours of labor on the part of the insurance company. All this costs money, which the insurer will pass along to beneficiaries in the form of higher premiums. Dealing with the employer's subscribers, on the other hand, is relatively simple. The insurance company can sell the coverage through just one negotiation. And it can handle the premiums and claims for all 500 beneficiaries through the employer's human resources department, which, if technologically competent, has systems in place to handle the job automatically.

Still, economics alone can't explain why the self-employed and people in small businesses (who face a similar situation) end up struggling to find affordable health insurance. The other part of the explanation lies in the history of American insurance—and the way the logic of competition ultimately unraveled the social bargain struck in the 1930s, when those early Blue Cross plans came into existence.

Although the Blues were private entities, they pledged themselves to a public purpose: bringing health insurance to large numbers of people. They did so by basing their premiums on a "community rate": every subscriber, regardless of age, sex, or medical condition, paid the same monthly amount. (Individuals paid higher rates than group members, because of the administrative overhead, but the difference was relatively modest.) Many insurers also practiced some form of "guaranteed issue," giving coverage to anybody who agreed to pay their premiums. These two features made Blue Cross resemble social insurance—a protection scheme in which healthy people subsidized the costs of the sick, precisely as the Committee on the Cost of Medicine had recommended in 1932, making insurance available to most people who had the money to pay for it.

While some pioneers in Blue Cross were legitimate do-gooders who saw the expansion of coverage as a crusade, the plans had a mandate to serve the interests of the hospitals that founded them. And the hospitals benefited from widespread insurance coverage, because it generated more paying customers and reduced the number of very sick charity cases on the wards. But it was the impact, not the rationale, that mattered to elected officials who blessed the Blue Cross plans with special charters, exempting them from taxes. These officials appreciated the fact that Blue Cross was helping so many of their constituents get access to medical care. And, in many cases, they also appreciated the way Blue Cross was undermining the case for national health insurance. Indeed, for the critics of universal health care, Blue Cross seemed like the best of both worlds: the benefits of government-financed medicine without the burden. The "brakes Blue Cross has applied to our swing towards socialization of medicine are perhaps the greatest benefits which we have all derived," a hospital administrator announced in 1950.

By that time, more than 20 million people had enrolled in Blue Cross, fueling predictions of near-universal coverage for the working population. But such predictions overlooked one crucial consequence of this growth—the signal it sent to the commercial insurance industry. Over the years, commercial insurers had at various times offered something called "sickness" benefits, generally as an optional supplement to their life and traveler's policies. Sickness insurance worked more or less as a disability policy would today: if you became ill and were unable to work, the policy would pay you some money to make up for your lost earnings. At first, these policies covered only disability related to certain types of diseases or injuries, such as those resulting from railway accidents. By the turn of the century, however, insurers had largely dropped such limitations, paying out benefits whenever somebody had a serious medical problem. A few years after that, insurers began paying out additional cash to those beneficiaries whose illnesses required hospitalization or surgery.

Sickness benefits worked well enough for the beneficiaries. But they were a nightmare to manage financially, because the people most interested in buying the coverage were precisely those who expected to need it, either because they already had serious medical problems or because they worked at jobs with a high likelihood of injury. In other words, sickness insurance was most attractive to the sick. This wreaked actuarial havoc, as too much benefit money was going out and too little premium money was coming in. When insurers responded by raising everybody's premiums to cover the cost of benefits, that made the problem worse, since the higher prices discouraged even more relatively healthy people from buying the coverage in the first place—ultimately leaving a group of beneficiaries who were in even worse relative health.

In the insurance business this phenomenon is known as adverse selection—the tendency of any voluntary insurance system to attract the greatest financial risks. And in the health care business, it was a daunting enough problem within the individual market to scare even large insurers like Prudential, whose president in the 1920s declared that sickness coverage could not "safely be transacted" by outside parties. But when Blue Cross demonstrated that health insurance really could work, just so long as it was based on large groups like employers, the commercial insurers decided to try again. By the 1950s, Prudential, Aetna, Metropolitan Life of New York, and the Connecticut General Life Insurance Company (which later became Cigna) all had thriving health divisions.

Their arrival signaled a critical change, however. Commercial insurers didn't have a mandate to serve the public good—or to fill hospital beds. Their goal was to make money. And the most obvious way to do that was to target groups of relatively healthy subscribers. Under the Blue Cross "community rate" scheme, these groups were, in effect, paying more to cover the expenses of less healthy beneficiaries. Commercial insurers offered to insure these people at rates that more closely reflected their own health status, then adjusted rates year after year according to how that status changed. This method of pricing insurance, known as experience rating, allowed commercial insurers to undercut the prices the Blues were offering.

Commercial insurers applied the same logic of selection to individuals who wanted to buy insurance—in no small part because these insurers knew, based on what had happened before the 1930s, just how prone the individual market was to attracting disproportionate numbers of high medical risks. And because they were dealing with potential customers one-on-one, they could be a lot more specific about whom to take, what kind of coverage to offer, and what price to charge. Adapting the knowledge gained in the life insurance business—life insurers had since replaced casualty insurers as the top purveyors of health coverage—they medically profiled each applicant, using such criteria as age, sex, occupation, and even race. Among the more bizarre factors insurers sometimes considered was height; apparently, extremely tall people were likely to incur higher medical costs. And when insurers felt that applicants posed inherently high risks, because of demographics or recent medical claims, they'd simply refuse coverage. For this very reason, the largest commercial insurers generally steered clear of the individual market altogether, figuring that the chance of attracting the highest-risk beneficiaries there was just too high.

The insurance companies didn't always boast about their selectivity. As a national survey of health insurance in 1961 explained, particularly when it came to controversial issues like race, they avoided "high-risk" clients—i.e., African-Americans and other minorities who were more likely to develop disease—simply by applying their other actuarial criteria more strictly to nonwhite applicants or soliciting business only from those customers where minorities were not present in large numbers. But the insurance industry made no excuses for its general approach of basing insurance availability and prices on risk. On the contrary, its leaders and allies explicitly rejected the egalitarian ethos that was the foundation of Europe's national health care plans and the early Blue Cross programs, saying the purpose of insurance was merely to spread the costs of expected medical care over time— and only within groups of people with a relatively similar likelihood of needing it. Besides, they said, community rating and other public service ideals were simply impractical in an insurance system not run by the government; and government administration of health insurance, they said, was simply un-American.

They were certainly right about the practicality of community rating in a nonuniversal, privately run insurance system. When enrollment in commercial insurance surged past enrollment in Blue Cross during the 1950s, the Blues finally began a slow retreat from their founding principles. First they began offering companies and unions experience-rated group contracts, just as the commercial carriers did, to meet the demands for cheaper coverage. And as they lost the ability to subsidize individual coverage with overpayment from group clients, they started abandoning community rating in the individual market as well, offering a range of premiums based on demographics and occupation. The creation of Medicare and Medicaid in the 1960s gave the Blues a reprieve, by removing from their portfolios some of the highest-risk individual beneficiaries they were carrying: the elderly and the very poor. But the trend toward more commercial behavior continued. By 1968, Odin Anderson, one of the country's leading experts on health insurance, declared that "the community rate concept is, for all practical purposes, dead."

Any lingering hope that private insurance might provide for everybody, including the highest-risk beneficiaries, died in the 1970s. As medical care became more expensive, the cost of insuring high-risk beneficiaries grew disproportionately, since those people used so much more medical care. Meanwhile, enactment of the Employment Retirement Income Security Act (ERISA) encouraged large companies to self-insure—in effect, to pay for their workers' medical claims on their own, out of company funds, instead of contributing premiums to large pools run by insurers. Given the relative youth and health of people working in large companies, that meant it would be even harder for insurers to subsidize the costs of the very sick.

These developments substantially raised premiums for small businesses, which didn't have the money to self-insure. But it was the self-employed and other individuals trying to buy coverage on their own who probably suffered the most—and who still have the toughest time finding affordable insurance today.

Indeed, the largest commercial insurers remain wary of the individual market, leaving it to the Blues and smaller, independent carriers. But except for the name and the logo, Blue Cross is a wholly different enterprise from what it was in the 1930s and 1940s. In 1986, Congress took away the tax breaks for Blue Cross, following a finding by the IRS that the plans were no longer very different from their commercial counterparts. This prompted many of the plans to convert to for-profit status outright. And although even these new, explicitly moneymaking enterprises have frequently professed a commitment to broad coverage in principle, many have adopted the same practices as the commercial carriers when it comes to individuals and small businesses—charging substantially higher premiums and screening applicants for prior medical conditions, refusing to cover preexisting conditions for months or even years.

The smaller commercial carriers are even more selective. Insurers wary of potentially risky clients sometimes demand statements from attending physicians, blood and urine samples, and even oral fluid tests. They have also been known to scrutinize applicants on the basis of class and income, for reasons that an official at Mutual of Omaha explained in 1999: "As an underwriter, you must be aware and ask why someone would spend 30 percent of their annual income to buy an individual major medical product, especially if they're 45 years old and have never had coverage before." In other words, it's a classic catch-22. Insurers serving the individual market assume that any low-income person willing to pay the ridiculously high prices being charged is probably too medically risky to insure anyway.

Insurers and others who defend these practices argue that failing to assess applicants with every bit of available information would leave them vulnerable to the same adverse selection problem that bedeviled the early sickness insurance policies and, much later, nearly destroyed the Blue Cross system. They are almost surely correct about that. Anytime an insurer has less information than either its applicants or its competitors, it is at a severe competitive disadvantage. The result, though, is that people with costly or serious medical conditions end up having a great deal of trouble finding affordable coverage. By doing what they must do to protect themselves, insurance companies are leaving the public—or, at least, the most medically needy portions of it—more vulnerable.

———

Diabetes mellitus, as it is known in the medical textbooks, is the inability to break down sugar. And it is a far more serious problem than such a pedestrian description might suggest. Sugar is as essential to human life as air and water. Within the chemical bonds linking its carbon, hydrogen, and oxygen atoms lies the energy that enables the body to function. A healthy person breaks those bonds apart—thereby releasing the energy—with the help of a hormone called insulin. But people with diabetes can't do this, either because their bodies don't produce enough insulin (that's type 1 diabetes) or because their bodies don't react appropriately to the presence of insulin (that's type 2 diabetes). The unstable sugar supply can weaken everything from the heart to the immune system, leading to any number of complications and even death.

Researchers still aren't certain exactly what causes diabetes. Type 1 diabetes appears in children. Type 2 has historically appeared mostly in adults, sometimes during old age, although lately it has also been showing up in children. Partly because it can appear so late in life, researchers believe that most cases of type 2 diabetes involve some kind of environmental "trigger"—such as eating a poor diet, getting too little exercise, and carrying around too much weight. (The sudden appearance in children seems to be related to growing rates of obesity in youth, particularly in low-income urban areas.) But research also suggests that even the adult-onset form of diabetes has a strong genetic component. In other words, some people are simply more predisposed to get it than others.

Once upon a time, diabetes killed most of those afflicted with it—quickly, in the case of those who developed it as children; or after years of suffering, in the case of those who acquired it later in life. Today, most diabetics can live a long, largely healthy life if they change their habits and take the right drugs—artificially injected insulin or pills that make the body's own insulin work better.

The development of these treatments for diabetes represents one of the greatest advances in twentieth-century medicine—but also one of the costliest. Individually, diabetics can pay thousands of dollars a year in medical costs, which include the supplies they use to monitor their blood sugar, their frequent doctor visits, and—particularly in their later years—hospitalizations from complications that will defy even the most conscientious preventive efforts. As a society, Americans spend more money on diabetes than any other single disease; this spending accounts for roughly $1 in every $10 flowing through the U.S. medical system.

Insurance companies know this. And although a little money spent on proper monitoring and treatment today can avoid a lot of money spent on hospitalization and rehabilitation tomorrow, insurers also know that beneficiaries change plans so frequently that the financial savings from proactive treatment might very well accrue to a competitor. This is particularly true in the individual and small-business insurance markets, where the churning between insurance plans is highest (both because people are more desperate for cheaper coverage and because people are less likely to stay with one employer for a long time). As a result, these insurers are the most likely to be on the lookout for diabetes—and the least eager to cover it.

Janice Ramsey had discovered this firsthand in 1999. Before that time, she had been relatively healthy. Except for childbirth, she had never even been in the hospital. But when she began complaining of fatigue and dryness in her mouth, her doctor suggested a "glucose tolerance" test—the standard diagnostic tool for diabetes, in which the patient drinks large quantities of supersweet fluids, then has his or her blood and urine tested for sugar levels at regularly spaced intervals. Sure enough, the test came back positive.

Janice was familiar with the disease, since her mother had it. And, perhaps because of that, she says, she took the news in stride. She knew that she could control the disease if she took care of herself and got regular medical care. But doing that would not be so easy. At the time of the diagnosis, Janice had just purchased a new individual health insurance policy, choosing it over the one she'd had previously because the premiums were a little cheaper. A few months later, the new carrier informed her, by letter, that it was dropping her coverage. Its rationale? It had determined that her diabetes was a preexisting condition—one that Janice failed to disclose on her application.

Janice says she had no idea what the insurer was talking about; she hadn't disclosed her diabetes, she says, because she didn't know she had it. But when she called to complain, the insurer told her there was nothing she could do. It turned out that this information was wrong: under the law in Florida, Janice had the right to challenge the insurer. (Like most states, Florida prohibits insurers that operate within its boundaries from retroactively canceling an insurance policy without good cause.) Janice didn't know that, however, and instead began her futile search for legitimate insurance coverage.

Janice kept no records of those applications, but her description of the experience appears to be typical. Several studies (including one published by the health insurance industry itself) have determined that, on average, companies tend to deny outright between 15 and 25 percent of the applications that make it to the underwriting stage. And many more applications never make it that far, because applicants give up once insurance agents or company representatives tell them that the chance of acquiring coverage is slim or nonexistent.

Some of the most illustrative evidence of this phenomenon was produced in 2001 by researchers from Georgetown University, who posed as seven fictitious people and applied for health insurance from sixty carriers across the country. "Alice," a twenty-four-year-old waitress, was healthy except for allergies she treated with a prescription drug. (She figured she couldn't do her job if she was sneezing all over her customers' food.) Just three insurers offered her "clean" coverage at standard rates. The rest charged extra premiums; imposed deductibles of between $500 and $2,500; or excluded coverage of allergies. A few actually refused coverage of upper respiratory problems altogether. "Denise" was a forty-eight-year-old actress and a survivor of breast cancer. Twenty-six insurers rejected her outright, and most of those that did offer coverage either had high premiums or excluded coverage of any future breast problems. The worst off was "Greg," a thirty-six-year-old writer who was HIV-positive. Not one insurer would cover him. Overall, only one-tenth of the applications submitted produced "clean" offers—that is, offers which included neither premium markups nor exclusions on coverage nor reduced benefits. "Consumers who are in less-than-perfect health clearly face significant barriers to obtaining health insurance coverage in the individual market," the authors of the study concluded. "Even applicants with mild health problems, or with a history of health problems that had been resolved years earlier, faced such barriers."

It mattered not a whit in these cases whether the applicants had money, or even a job. And this, more than anything else, was what bothered Janice. She felt that she had paid her dues—and then some. Yet whenever she had a serious medical problem, she found herself in the emergency room, begging for charity. Anger turned to shame and then to desperation—desperation that made her the perfect target for a swindle.

———

One day in the late 1940s, a man riding the subway in New York City broke his leg when his car stopped suddenly, hurtling the passengers forward. He had health insurance through a small carrier and filed a claim, thinking the injury qualified for coverage. But this particular policy, according to an account in the *New York Times*, turned out to contain some curious fine print, including a clause stipulating that the insurer would pay benefits for injuries incurred during railroad accidents "only if the common carrier"—in other words, the railroad car—"was damaged, too." In this case the subway car didn't suffer any damage; it had just stopped short. So the company refused to pay.

Bewildered, the man eventually complained to state officials. And so, apparently, did more than 500 other people who discovered similarly obscure exceptions to their coverage hidden deep within their insurance contracts. The New York State Department of Insurance investigated and, eventually, took legal action against more than 100 companies that, according to the state, had defrauded New Yorkers of some $10 million in premiums over the course of a year by deliberately misleading customers about benefits.

As the department's investigators found, the companies' typical modus operandi was to deploy slick salesman brandishing professional-looking literature and promising comprehensive benefits, all at bargain-basement rates. But when the holders of these policies filed claims, they frequently learned that the contracts actually restricted coverage to only a small, very rare class of diseases and accidents. This enabled the bogus insurers to keep most of the premiums. In some cases, the state's officials found, the companies had paid less than 10 percent of the premium income as benefits.

These "gyp" artists, as the newspapers called them, did most of their damage in the city's low-income, heavily African-American sections and in some of the poorer, predominantly agricultural communities upstate. Those areas had high concentrations of people shopping for insurance on their own, since they had no access to the job-based group health insurance. More crucially, many lacked either the expertise or—because of their long work hours—the time and savvy to distinguish a real policy from a phony one.

About thirty years later, a new generation of scam artists was engaging in virtually the same kind of schemes, with one twist. Instead of focusing on the least affluent, least sophisticated neighborhoods of New York, the new scam artists were focusing on the least affluent, least sophisticated corner of the private health insurance market: individuals and small businesses. Once again, "insurers" typically approached their clients by offering generous benefits and unusually low rates, all without the intensive medical underwriting and exclusions for preexisting conditions so familiar to people who'd tried unsuccessfully to buy affordable insurance. The companies claimed to be offering their coverage through unions, professional associations, or groups of employers from different states; by forming larger groups, they explained, they were able to get the same kinds of prices and benefits as large employers.

The plans would typically pay small claims for a while, so that beneficiaries kept paying premiums; eventually, though, big claims would roll in and the plans would refuse to pay them. State authorities would eventually catch up with the plans and shut them down, but usually only after they'd extracted months' worth of premiums. One bogus plan, promoting itself as a union plan organized by the nonexistent "International Worker's Guild," left some 31,000 people owing about $25 million in outstanding medical bills. In the wave of scams that hit between 1988 and 1991, almost 400,000 people were left owing more than $123 million in unpaid medical bills.

A key ingredient in the success of these schemes was widespread confusion over which authorities were responsible for regulating private insurance. Historically this responsibility had belonged to the states. But after 1974, when ERISA passed, states lost the authority to regulate self-insured plans like the ones that large corporations were increasingly offering.

This development suited large businesses just fine, since they had grown tired of tailoring their employee insurance plans to the varying requirements of different states. But the law had some perverse, unintended effects—among them, creating considerable misunderstanding over who would be responsible for regulating health insurance offered through groups of professional associations or small businesses that had decided to band together and buy insurance as conglomerates. Previously, agents and potential beneficiaries could spot a scam by demanding to see an insurers' state license. Once ERISA became law, though, scam artists could claim that they were exempt, and even intelligent consumers might not know any better.

The scam subsided during the mid-1990s—once health insurance briefly became more affordable again, state officials began looking out for plans falsely claiming ERISA exemption and Congress finally clarified the law's intent. (Previously, state officials weren't entirely certain whether they had the authority to shut down fraudulent insurers if they operated through groups; ERISA theoretically could have meant it was a federal responsibility.) But not long after premiums started rising, toward the end of the decade, new schemes started popping up, as a new breed of even more sophisticated con artists began taking advantage of the renewed need for more affordable coverage. (One fifty-year-old woman taken in by a scam recalled being offered monthly premiums of $285; the cheapest premium she could find elsewhere was $425.)

In addition to using promotional material that looked authentic, complete with fake papers verifying the exemption from state laws, many of the plans also secured real contracts with existing networks of doctors and hospitals, just as authentic insurance plans were doing, thereby giving the bogus benefits a veneer of legitimacy. Some, like an unlicensed plan called Employers Mutual, actually marketed their policies through existing, well-respected professional associations. (Employers Mutual sold some of its policies through the National Writers Union, an organization for professional journalists.) Others, like the "Vanguard Asset Group," adopted names that sounded like well-known financial institutions. (The Vanguard Group is one of America's best known investment houses.) And although scam artists popped up all over the country, they concentrated most of their efforts in the south and west, where the relatively high proportion of small start-up businesses created a target-rich environment.

The insurance plans sold most of their coverage to individuals through licensed agents. And in at least a few cases, agents familiar with the ins and outs of ERISA were suspicious enough about the coverage to make inquiries with state insurance regulators, tipping them off to the scam. But for the most part agents didn't question the legitimacy of insurers like American Benefit Plans—either because they simply didn't spot the warning signs or because they were too distracted by the unusually large commissions the phony plans paid. (In a few cases, the agents themselves were part of the scam.) In those instances, the plans were once again able to buy time for themselves by paying out small claims, then explaining to beneficiaries that they were delaying payment on the big claims only because of billing disputes with the hospitals. By the time state regulators started hearing from frustrated patients, the bogus insurers had done plenty of damage. Even in those instances when officials were able to shut down the fake plans immediately and seize their assets, the money recovered was just a fraction of the outstanding claims. Between 2000 and 2002, according to a study by the General Accounting Office, 144 different "insurers" sold policies to some 200,000 policyholders, leaving them with $252 million in unpaid claims. Officials were able to recover just 20 percent of that money by 2003, the study said, with little hope of getting more.

Florida was the unofficial ground zero for these scams. In Deerfield Beach, Richard Baer, the owner of a small greeting card shop, had no idea that his insurance from "TRG Marketing" was faulty—until, like Janice Ramsey, he started getting a series of unpaid bills from doctors and hospitals eventually totaling about $60,000. The state closed TRG shortly thereafter, indicting its two founders. (In 2005, they would plead guilty to charges of racketeering and unlawful insurance transactions and agree to serve terms of two and four years, respectively, followed by 20 years of supervised probation.) But state officials were not able to extract any money from the scam.

Baer, an army veteran who'd spent twenty-five years as a salesman in New York before moving to Florida, managed to negotiate reduced charges from most of his health care providers and pay the bills out of pocket. But the ordeal left him $20,000 poorer and without health insurance, since his history of recent heart problems plus his age, sixty-one, made him effectively uninsurable at anything close to affordable rates. (Part of the problem was that he needed what money he had to pay for insurance for his wife, since her medical problems were even worse than his.) He would remain uninsured until early 2006, when he finally turned sixty-five and became eligible for Medicare.

In Okeechobee, a rural town just north of the giant lake by the same name, Rusty Baker had also purchased coverage from TRG. According to an account in the *Okeechobee News*, the coverage was supposed to be for him and his wife, Cindy; they ran a home roofing business together. The Bakers had switched from Humana in order to shave about $100 off of their monthly premiums. But then Cindy incurred two substantial medical charges: a hysterectomy, and radiation therapy for skin cancer. (The two were unrelated.) TRG never paid those bills, and the Bakers, like the Baers, soon learned that the company was a fraud. That left the Bakers owing $60,000—and uninsured for future expenses. Rusty, who'd been battling an addiction to painkillers after a series of on-the-job back injuries, ended up taking his own life; Cindy, who had difficulty obtaining coverage for her preexisting medical conditions, said the bills were what pushed him over the edge.

Perhaps the most famous victim of an insurance scam in Florida was Pete Orr, a popular stock car driver who, along with his wife Teri, had opened a racing school near Orlando. Thanks in no small part to the risks of the racing, they had a hard time finding affordable coverage—until they discovered Employers Mutual. The Orrs paid their premiums on time, and for a while Employers Mutual paid their claims. But then Pete was diagnosed with lymphoma, and it was the same old story all over again: once the bills got to be large, Employers Mutual stopped paying them. By the time the Orrs figured out that their coverage was bogus, they already owed more than $100,000 for Pete's chemotherapy—and, of more immediate concern, they had no insurance for new procedures. As Pete's condition deteriorated, doctors recommended a bone marrow transplant; Pete's brother, it turned out, was a perfect match. But the procedure would have cost $300,000, and the hospital wouldn't do it until the Orrs came up with at least some of the money. The Orrs finally put together the funds—thanks, in part, to a telethon on Pete Orr's behalf—but by then several months had passed. Pete died before the procedure could be done. All Teri had left were the medical bills, amounting to about $285,000 in all.

In the immediate sense, Janice Ramsey was a little more fortunate. She was still alive, for one thing. And the bills she and her husband owed came to $20,000 in outstanding medical charges—a relatively modest sum compared with what some other people in the state owed. But they had let bills go unpaid during those few months when American Benefits claimed it was still legitimate. It would take them time to pay off those debts, since they had little savings. (As home builders, they'd put most of their equity into their house.) And Janice says her credit rating suffered accordingly. Before she lost her insurance and began to struggle with her medical bills, her credit score was 780, or nearly perfect. Afterward, it was down to 630. That meant she was no longer eligible for the best rates on loans. (Before her medical debt problems, she liked to point out, she had even qualified for interest-free financing on a Chevy Suburban.)

More important, though, Ramsey was now uninsured all over again—with nobody particularly eager to pick her up. "There's nothing out there to help me take care of myself, and that's all I'm asking," she would say later. "I'm not asking for anybody to give me anything, just let me be able to buy it." Medicare and Medicaid weren't options, either. She wasn't old enough to qualify for the former, and she still wasn't poor enough to qualify for the latter. (Her husband, on the other hand, was already sixty-five, so he did qualify for Medicare.)

At one point Janice heard about the Health Insurance Portability and Accountability Act (HIPAA)—a relatively new federal law designed, in part, to help people with serious medical conditions find affordable health insurance. The law was considered a breakthrough when it passed in 1996. But it was utterly useless to Janice. To qualify for its benefits, you had to have prior coverage under either a large group plan or an individual plan licensed in the state. But Janice had neither: she'd bought the insurance on her own and American Benefit, of course, had no license. So, perversely, the very same problems that made her uninsured in the first place now made her ineligible for assistance under the federal law supposedly designed to help her. (Even if Janice had qualified, it's not sure whether she would have been able to afford HIPAA insurance, since the law of 1996 didn't regulate how much insurers could charge for their coverage. One subsequent government study found insurers hiking premiums as much as 2,000 percent for applicants with prior medical conditions.)

Next Janice turned to the state of Florida—which, at first, seemed like a more promising option. Like many states, Florida had debated and passed a series of incremental insurance reforms after the failure of Clinton's health care plan in 1994. Several of these reforms were designed to make coverage more widely available to small businesses and people purchasing coverage on their own, since, historically, those groups had so much trouble. One innovation was to create a "group of one" market in which a single self-employed individual could apply for coverage during a one-month period every year. During this window of opportunity, designed to mimic the "open enrollment" periods large employers gave their workers, private insurers that operated in the state agreed to provide coverage to anybody who wanted it. In other words, the private insurers were offering "guaranteed issue," just as the Blue Cross plans once did.

But the insurers doing business in Florida agreed to do this only under certain conditions that would protect their financial liability. In particular, they reserved the right to exclude coverage of preexisting conditions. In other words, even though they agreed not to deny applications outright, they insisted on the right to screen applicants in other ways. And in Janice's case they did just that. The best policy she could find had monthly premiums of about $700—with a $3,000 deductible up front. It also had a two-year "exclusion" period: during that time, it would cover nothing related to her diabetes—not her medical supplies, not her regular checkups or tests, and most certainly not treatments for complications.

In the end, Janice figured, it came down to simple math. She could spend about $17,000 over two years for insurance coverage that would cover nothing related to her diabetes. After the two years, she could continue to pay $700 a month in premiums—or more than that, since the rates could go up—for coverage that would still leave her responsible for the $3,000 annual deductible plus cost-sharing for each doctor visit, test, prescription, and hospitalization. (Not to mention the $300 a month she spent on testing supplies, artificial tears, and such.) Or she could go uninsured and hope for the best, figuring that at least she'd have the extra money, with which she could start to pay off any catastrophic bills. Still not entirely confident that the insurance would really be there two years later anyway—hadn't American Benefit seemed perfectly legitimate?—she chose to take her chances without insurance.

The choice had consequences. In late 2003, a dull pain in her back grew until she could no longer sit down in a chair for extended periods of time, making it virtually impossible to do her job. Eventually doctors found a tumor. The good news was that the tumor was not malignant; although painful, it was not life-threatening. The bad news? Precisely because it was not an emergency condition, she'd have to pay for the operation herself—at a price of several thousand dollars. It took her nearly a year, until she found yet another state insurance program, this one for work-related disabilities, for which she was eligible.

Another problem for Janice was that she was still working, as a consultant, and making decent money. The public clinics in her community, which provided free care for destitute residents, wouldn't take her because she wasn't poor enough for their eligibility guidelines.

Doctors and hospitals discounted their fees sometimes. But they still demanded substantial out-of-pocket payments; and the hospitals would take care of her only in true emergencies. "Every three months you're supposed to go for blood work to see where your diabetes is at," she said at one point. "I go twice a year because it's $400 each time you go. Nobody can pay for that. My prescriptions nobody pays for either, you know. . . . I don't understand why our health insurance is so poorly run. It really is. It seems like 'hooray for me. I know you're sick, too bad. I don't care. It don't matter to me.'"

In 2005, Janice got a realtor's license. And that allowed her, finally, to obtain legitimate health insurance once again, through a bank that offered coverage to realtors who worked with it. She wasn't sure how long she would be able to keep this insurance, since in order to qualify she had to sell three mortgages a year. Still, it was group-style coverage with no exclusion of preexisting conditions. For that, Janice was thankful.

But Janice's eventual success at finding coverage may come to seem even more anachronistic in coming years. If group health insurance continues to erode, more people will have to search for benefits on their own. And their search is likely to become increasingly difficult for several reasons, starting with one of the great medical advances of our time: the unlocking of the human genome.

In the last few years, scientists have begun learning not only to identify which medical conditions have genetic origins, but to identify the specific genes that predispose a person to those conditions. Probably the most famous example of this is the discovery of the so-called breast cancer genes. Researchers pinpointed two of these, BRCA I and II, during the 1990s. When the genes have the correct coding, the enzymes fix cells that are broken and keep cells from reproducing too quickly. But when the genes themselves are damaged, they can't produce the right enzyme. That allows cells to grow out of control—in other words, to become tumors.

Women who have mutations in these genes have greater than a 50 percent chance of developing breast cancer. As a result, women who suspect they may be at risk (typically because of a family history) routinely test for the presence of the mutations. Those who have the errant genes may opt merely to be more vigilant about screening and self-examinations, or they may choose preemptive surgery. Oncologists, in turn, now base therapy on which cancer gene a woman has. And someday, perhaps soon, scientists could tailor chemotherapy treatments even more individually, identifying not just which gene is defective but what kind of defect it contains.

Other cancers, Alzheimer's disease, heart disease, and, yes, diabetes—all these diseases seem to have some genetic link. If and when scientists discover them, testing and treatments will both improve, producing yet another leap forward in medicine and, ultimately, in health. In fact, one of the fastest-growing and most intriguing fields of medical research today is pharmacogenetics, which is the study of how a person's genetic idiosyncrasies affect the way that person reacts to different medications.

But as genetic testing improves, doctors and patients won't be the only ones interested in the results. Insurers will be interested, too, because it will help them identify, in advance, which people are at the greatest risk of contracting serious diseases requiring expensive treatment. "The insurance industry's great fear is that genetic testing will become the norm in clinical practice . . . and that insurers will be denied access to this newer information," one official at an insurance company explained a few years ago. "Genetic screening may become so common in the future that the insurance industry feels compelled to start performing the same type of testing."

And even if genetic testing doesn't prove as revealing as experts expect, other technological developments will help insurers to assess potential beneficiaries more carefully in the future. Already, virtually all insurers in the individual and small-business markets cross-reference applications with a national clearinghouse called the Medical Information Bureau, which as of 2005 had the past application records of 30 million people who had applied for either life or health insurance in the previous seven years. More recently, pharmacy benefit managers, the private firms that negotiate discounts on prescription drugs for large and small insurance companies, have begun making their records available for the same purpose—data that apparently yields information for about one-third of all applicants in the individual market, a proportion that will probably get higher as the information becomes more sophisticated.

At the moment, insurers' ability to actually use such information is still somewhat crude. But that could change with the development of a mathematical technique called predictive modeling, which produces projections of future medical expenses based more closely on a specific person's medical history. Traditionally, actuaries have based rates and coverage exclusions primarily on fairly crude estimates taken from large population samples: the average diabetic runs up so many dollars in expenses a year, so somebody with diabetes should be charged so many dollars extra a year in premiums; and so on. With predictive modeling, actuaries can plug figures from an individual's recent billing history into an algorithm, then produce an estimate of the expenses that person is likely to incur the following year. It's a new science, and some experts question how accurate the models can be. But if they prove useful, insurers will have a new and potent tool for assessing their financial risk beforehand—and adjusting what coverage they offer accordingly.

These are just some of the reasons the insurance industry could become even more selective in the coming years, thereby making it harder for people with serious medical conditions to buy affordable coverage on their own. And while states over the years have passed regulations to help high-risk individuals get coverage, those rules are clearly insufficient to help people struggling to find coverage today— let alone those who might be trying tomorrow. The effects would be medical as well as financial. Already, according to one study, some women with family histories of breast cancer are deciding against taking the genetic test because they fear a positive result might make it harder to find insurance in the future.

Yet rather than trying to bolster these protections, President Bush and the Republicans have spent the last few years trying to weaken them—in effect, giving insurers more leeway, not less, when it comes to weeding out the medically riskiest customers. They have justified this effort by arguing that existing regulations—not just on rating and pricing practices, but on specific benefits as well—load the individual insurance market with so many burdens that the policies inevitably become too costly for many people to afford, thereby driving up the number of people without insurance. Their proposed solution: Allowing individuals and small businesses to buy insurance across state lines or in groups subject only to relatively lax federal guidelines; either way, many, if not most, existing state regulations on the behavior of insurance companies would become effectively irrelevant.

It's certainly true that regulation drives up the cost of benefits, and that this ultimately means fewer people will buy insurance (either out of choice or out of necessity). Existing regulations are, at best, a highly imperfect solution to the problems of the individual insurance market. But if either of these measures made health insurance nominally cheaper, it would do so only by making it less available to people with serious medical conditions—the ones now protected, however inadequately, by the existing regulations. In addition, creating a vast, nationwide market for individual health insurance would probably lead to even more frauds like the one that snared Janice Ramsey.

Ultimately, the problem with the individual health insurance market is the nature of the private health insurance industry. When they're not dealing with large groups of employees, insurers have no desire to protect those who most need protection. This reluctance is understandable, even sensible, given the history of the industry and the realities of competition. It just so happens that such behavior is not always in the best interests of the people buying insurance—particularly those, like Janice Ramsey, who need it most.

THREE

AUSTIN

The city of Lakeway sits on the southern banks of Lake Travis, a sixty-four-mile-long reservoir that winds through the hill country west of Austin, Texas. For most of its history, Lakeway was more a place to visit than a place to live—a mere destination for city folk hoping to boat, fish, or swim on what the Texas Water Commission once declared the state's clearest reservoir. But that started to change in the 1980s, when Dell Computers established its headquarters in nearby Round Rock, transforming Austin and the surrounding communities into a sprawling, high-tech metropolis bursting with young, well-educated professionals. It didn't take long for these new arrivals to find Lakeway and its large swaths of still undeveloped land. By 1990, the year-round population was swelling past 4,000 people, more than quadruple what it had been a decade before, with many of the new residents settling into upscale gated communities springing up on the ridge between the city's lush golf course and lakeside resorts. Lakeway's reputation for insulated suburban bliss soon earned it the moniker "Fakeway"—the kind of place where, it seemed, the only worry in life was what to do about the stubborn deer population unwilling to make way for the new three-car garages and swimming pools.

In 1991, Elizabeth Hilsabeck craved the Lakeway lifestyle, in no small part because it was right under her nose. She and her husband, Steven, had been renting a duplex in one of the town's older sections. And they knew it would be hard to find a house in Lakeway they could afford. While they both had well-paid jobs—he at a bank, she at a company that outfitted booths for corporate trade shows—they were just in their early twenties, with only a little money saved between them. But fortune smiled on them. One night their realtor called with a new listing in Lakeway. And it was in their price range.

Elizabeth and Steven arranged to see it the very next day, and it didn't take long for them to realize why the three-story, split-level home was so affordable. It was an older structure with decor straight out of *The Brady Bunch*: green shag carpet, neon orange trim on the kitchen fixtures, and hanging beads in some of the doorways. The overgrown lawn was full of weeds, and although the house itself was clean, the whole place reeked of cigarettes.

Still, the home had potential. Its 2,400 square feet included three bedrooms, perfect for a young couple, plus a top floor whose windows revealed almost 360-degree views. It had room for a two-story deck, offering a glimpse of the lake. And it was in a great location, on a cul-de-sac that backed onto the Lakeway airstrip—a quiet private airport used by locals who owned their own planes. (Some of the neighborhood's larger houses actually had taxiways that led right up to the back door.) The Hilsabecks made an offer right away, and the owners, apparently eager to sell because of an impending divorce, accepted.

Over the next few months, Elizabeth and Steven methodically set about realizing their vision. They ripped out the old carpet and wallpaper and painted most of the house white. In the living room, they installed a wood stove to replace an old stand-alone fireplace, then decorated the whole area to look like a mountain lodge. Steven got to work on his deck, using material from the old porch, and they remodeled the kitchen, complete with new countertops and a pass-through for entertaining. On the long, warm summer nights, the two would sit on their lawn chairs, nursing strawberry daiquiris and watching the single-propeller planes float down against the tangerine dusk. "We worked side by side; we were best friends," Elizabeth would say later. "It was the best time of my entire life."

The only thing missing from their lives was children. And that, too, was about to change. The following spring, Elizabeth became pregnant. A few weeks later, a doctor examining her thought he heard more than one heartbeat. When a sonogram confirmed that the Hilsabecks were having twins, Elizabeth and Steven began to busy themselves with preparations, mulling over names and making plans for decorating the nursery. Native Texans both, they finally decided they would go with a southwestern theme: peach and green, with wallpaper that featured cactus trees, bunnies, and dancing coyotes.

Everything seemed to be just fine—until one day in November, when Elizabeth began to feel a series of sharp pains. For three days her doctor prescribed Tylenol and rest, figuring that she was having so-called Braxton Hicks contractions (the contractions of early pregnancy, which the body uses to build up strength for the eventual delivery). But when the pain intensified, the doctor brought her into the hospital, where an exam and some tests confirmed her worst fear: she was in premature labor.

For the next twenty-four hours, the staff kept Elizabeth virtually immobile and under close observation, in an effort to arrest the delivery. But Elizabeth's small body—her five-foot-two frame had carried just 100 pounds before the pregnancy—wasn't going to hold the twins any longer. As her obstetrician started making preparations for a cesarean section, a neonatologist sat at Elizabeth's bedside to prepare her for what was likely to happen. Survival for premature babies was likely only after twenty-six weeks of pregnancy, he explained, and Elizabeth had just finished her twenty-fifth. The next morning, the babies were born—a girl the Hilsabecks would name Sarah and a boy they would call Parker. Although Elizabeth was nearly delirious from the ordeal, she managed to give each baby a gentle kiss on the forehead. Then she slipped back into a woozy trance of exhaustion and anesthesia, as the nurses whisked the newborns off to the intensive care unit.

The immediate prognosis was as grim as predicted. Each infant weighed just one pound, thirteen ounces—small enough to fit in the palm of an adult hand. Both needed breathing and feeding tubes, in part to treat "respiratory distress syndrome"—an ailment in which the underdeveloped air sacs inside the lungs of premature infants can't expand fully to absorb oxygen. In addition, Sarah's heart tended to stop when her body relaxed—another common problem among such premature babies—so doctors had placed her on an artificial pillow that was constantly deflating and reinflating, to keep her from falling too deeply asleep. Before Elizabeth went to see the babies in the nursery for the first time, the nurses had prepared her by showing her Polaroid photos, labeled "Hilsabeck boy" and "Hilsabeck girl." "I had no idea something so little could be alive," she would later recall. And even with this preparation, Elizabeth still found it difficult to see her babies, pinned down under so much equipment, then separated from her touch by a plastic covering. That night, when she was back in her room, she overheard a new mother, from down the hall, asking nurses to take her baby boy away because she needed sleep and he wouldn't stop crying. All Elizabeth could think was that her babies couldn't cry—because they had tubes stuck down their throats.

Still, Elizabeth knew that all the gadgetry served a purpose: keeping the babies alive. And despite all the warnings about how unlikely the children were to survive, she held out hope, counting every gram, day after day, as they slowly gained weight. After the respirators came out, the babies made their first sounds—more whispers than cries, but sounds all the same. A miracle, Elizabeth thought. At about a month and a half, they opened their eyes. Neither infant was out of the woods by any stretch, but now the doctors began to talk more openly about the possibility that both might make it after all. The doctors were careful not to promise too much; even if the children lived, they warned, there was a high risk of severe mental and physical disabilities. But Elizabeth didn't care. One day she found herself kneeling in front of her wedding picture, which featured a large cross in the background. Although she wasn't particularly religious, she prayed: "I didn't care if they were deaf, dumb, blind, or crippled. I just wanted them to live." They did. They had defied the odds. And in March 1993, after four months in the hospital, they came home for the first time.

By this point, Elizabeth's motherhood had long since ceased to be ordinary. Because of the premature births, she missed out on the baby showers and other familiar rites of expectant mothers. And now, because of the twins' health problems, she had to rearrange her entire life in order to accommodate their needs. Elizabeth had been taking night classes, in order to finish the college education she'd interrupted to marry Steven. She dropped those. She also decided to quit her job, in order to become her children's full-time attendant. Her doctors had told her to monitor the babies' progress closely, and she took the advice literally, creating giant spreadsheets, to track the children's food intake, medications, and weight gains. Later, she would learn how to administer intravenous (IV) medication.

But there was one challenge for which the doctors hadn't prepared Elizabeth: a struggle over health insurance. Her problem wasn't that she lacked coverage, as Gary Rotzler and Janice Ramsey did. The

Hilsabecks had health benefits through Steven's job at the bank. And, as far as they understood, it was excellent insurance provided by one of the best-known companies in the insurance business. But this coverage was through a health maintenance organization, or HMO. Unlike the traditional insurance to which they and most working Americans had long been accustomed, the HMO would pay for bills only if the Hilsabecks stayed within a preapproved network of doctors and hospitals. The HMO also scrutinized treatment, refusing to cover those services it deemed unnecessary, even if that decision contradicted a doctor's recommendation. This is exactly what would happen to the Hilsabecks, provoking an extended fight that would come to dominate their lives in the next few years.

The Hilsabecks, of course, were not the only ones fighting their insurance carriers. HMOs were a form of "managed care," a type of health insurance in which insurance companies demand lower prices from health care providers and intervene directly in medical decision making. By 1993, the year after Parker and Sarah were born, 51 percent of all workers with employer-sponsored health insurance belonged to some form of managed care—up from just 10 percent a decade before.

Managed care had become popular because businesses believed it could control the rising cost of health insurance, restoring some much-needed sanity to the world of medical finance. And by that criterion, managed care was a success. During the 1990s, the cost of health care rose more slowly than at any other time in recent memory. At one point, health care inflation actually fell below the rate of general inflation—which is a fancy way of saying that the price of medical care was increasing more slowly than the price of other goods.

Because few Americans were accustomed to restrictions on their access to doctors or treatments, managed care was always less popular with consumers than it was with employers—a status the promoters of managed care insisted was undeserved. As they correctly pointed out, most people in HMOs ultimately reaped benefits in the form of more affordable premiums, more comprehensive benefits, or larger paychecks. At a time when employer-based health insurance was becoming ever more precarious, managed care could truly be said to make coverage more secure. And that was good for everybody.

Or was it? The basic premise of managed care was that sound economics would also produce sound medicine—that cost control as practiced by HMOs and encouraged by employers would improve the quality of health care even as it saved money. But for families like the Hilsabecks, whose members needed costly treatment for chronic medical conditions, it didn't work out so neatly. And as the years went on, the Hilsabecks began to wonder whether this new health insurance system was designed to put the wonders of American medicine within reach, or to push them farther away.

Managed care has become an object of such scorn in contemporary America that it's easy to forget its idealistic lineage, which goes back, at least in part, to Depression-era rural populism and an immigrant doctor named Michael Shadid. In 1928, Shadid decided he wanted to give something back to Elk City, Oklahoma, the farming town that had embraced him when he first immigrated from the Middle East some two decades before. Like most of rural America, the farmers in Elk City had trouble getting medical care: doctors were hard to find, and their fees were hard to afford. To remedy this, Shadid proposed to create a cooperative. The families would pay regular, fixed fees to the cooperative; Shadid would then use the money to hire some doctors and set up a clinic, which would provide the families with comprehensive medical care.

As the sociologist Paul Starr notes in his history of American medicine, this was a radical idea. Doctors had traditionally worked on a

fee-for-service basis: they charged—and were paid—for every different service they provided. Shadid was proposing to put his doctors on
a salary. And Shadid's clinic would operate as a true cooperative, with
the people who paid money into it electing representatives to a board
which would have final authority over the clinic's administrative and
financial operations. Although Shadid deliberately preserved the physicians' authority over medical matters, his local colleagues perceived
the co-op as a threat to physicians' cherished autonomy—and reacted
accordingly. They refused invitations to join the co-op practice and
told would-be recruits from other communities that the plan was "in
disrepute." Eventually the physicians of Elk City even tried to revoke
Shadid's license, going so far as to dissolve the medical society and
reconstitute it in order to exclude him from membership. (They worried that simply expelling him might violate the existing charter.) Only
the support of the Oklahoma Farmers Union and other local political
groups allowed the project to go forward—and, eventually, thrive, in-
spiring dozens of such co-ops across the heartland in years to come.

At roughly the same time that Shadid's clinic was opening in Oklahoma, a similar model for medical care was establishing a foothold
half a country away. In 1929, two doctors in Los Angeles signed a
contract to provide comprehensive medical care to employees of the
county water department. In 1933, several more doctors in Los Angeles agreed to form a group practice that would provide health care for
workers on an aqueduct construction project. The project was being
run by the industrialist Henry J. Kaiser, who was so taken with the
success of the medical group that he made similar arrangements for
his workers on Grand Coulee Dam in Washington state and later at
his West Coast shipyards. When World War II ended, Kaiser opened
those plans to the public instead of dissolving them, forming the first
"Kaiser Permanente" health systems.

Prepaid group practices, as they became known, started cropping
up in other places, too: Chicago, Minneapolis, Seattle, Saint Louis.

Although not identical, they shared some defining characteristics reminiscent of Shadid's clinic: instead of charging patients for each service rendered, they instead asked members to pay fixed monthly fees. Doctors worked primarily for a salary rather than on a fee-for-service basis, and tended to collaborate more than physicians in private practices. The organizations themselves were officially nonprofit, meaning (at least in theory) that surpluses were plowed right back into patient services. And in some cases patients were involved in managing the groups, typically by electing representatives to a board with decision-making power.

The American Medical Association (AMA) greeted this development with the same hostility Elk City's physicians had unleashed on Shadid, convincing many states to effectively outlaw group practices by regulation and by blacklisting physicians who participated. But the group practices proved popular with their patients, who appreciated both the affordable benefits and the convenience of a single "medical home." They also captured the attention of a pediatric neurologist named Paul Ellwood. A California native, Ellwood had come east to Minneapolis during the 1950s to run the Sister Kenny Institute, a polio hospital. When Jonah Salk's vaccine rendered that work obsolete, the hospital began focusing on other rehabilitative services. And that's when the odd reality of medical financing occurred to him. As Ellwood tells the story, the clinic's doctors had gotten very good at making their patients better, shuttling them out of the hospital ever more quickly. Yet fewer days in the hospital meant fewer charges to the insurance companies—and, therefore, less income. The Sister Kenny Institute was nonprofit, but it still needed to cover its expenses. And it was having a great deal of trouble doing so. "Ellwood was faced with both a practical and [a] moral dilemma," Joseph Falkson, a historian of managed care, wrote in 1980. "How could Sister Kenny Institute continue its 'good works' when the very success of those efforts would ultimately lead to bankruptcy?"

Ellwood thought there were better ways to run a hospital—and health insurance. He was familiar with the Kaiser California plans. And since coming to Minneapolis he had seen up close the nearby Mayo Clinic, a group practice widely considered among the world's finest medical establishments. The patients in these group practices got better preventive care—through immunizations, regular checkups, and early tests—as well as more effective treatment for disease than patients in traditional fee-for-service health insurance plans. The reason, Ellwood surmised, was that the incentives rewarded quality, not intensity, of treatment, while the group model fostered better communication and cooperation among physicians in different specialties.

Ellwood began talking up group practices, and in 1970 he got the attention of the Nixon administration, which was desperate both for a way to control the suddenly skyrocketing cost of medicine and for a way to stop Senator Ted Kennedy's new push for national health insurance. At a meeting in a Washington hotel room, Ellwood laid out his vision. Rather than have the government take over the insurance business, as liberals like Kennedy always talked about doing, government could simply induce the private sector to create group practices. By providing superior care at lower cost, the group practices would lure beneficiaries away from traditional insurers and eventually compete among themselves, making medicine more affordable—and better—for everybody. The market, in other words, would save the American health care system from its own excesses. At one point, one of Nixon's advisers—clearly intrigued—asked if there was a name for these sorts of group practices. Ellwood figured that the first priority of these entities was to keep people healthy. Call them "health maintenance organizations," he said.

Now this was the kind of initiative a Republican administration like Nixon's could get behind. It was based on voluntary participation: Employees would have incentives to join HMOs, but they would be free to refuse. Better still, an HMO system wouldn't control costs by having the government set prices, as universal health care systems typically did; it would simply rearrange economic incentives to fit the unusual nature of medical care. Still, Nixon's advisers did propose some important twists. In fostering the creation of HMOs, by putting up some money and establishing accreditation standards, Ellwood had hoped the new entities would look like Kaiser and Mayo—that is, they'd be nonprofit group practices. Nixon's advisers, wary of overly specific government regulations, had other ideas. Why define HMOs so narrowly? Maybe investors would see this as a lucrative opportunity, thereby making much more capital available. And maybe, Nixon's advisers thought, not all the doctors had to work together on salary. Maybe some could just contract with insurance companies, working at discounted rates but still within the old fee-for-service structure. The incentives would still favor thrift, the argument went, even if the organizational architecture was a little different. Ellwood surmised that the administration's preferences reflected both ideology (a belief that more wide-open competition would produce a better result) and practical politics (focusing too relentlessly on group practices would inflame the physician community). But he understood that thinking— and, in many ways, agreed with it. So he went along.

After Nixon signed the HMO Act in 1973, Ellwood became the most visible promoter of the concept, laying out his idea to business audiences in the hope that large employers would encourage their workers to enroll and, then, push HMOs to focus on quality. He finally started to make headway in the 1980s, when the long-standing political alliance between corporate America and organized medicine began to rupture. With the costs of employee benefit plans explod-ing, businesses came to see doctors and hospitals, who had virtually unlimited power to set fees, as their adversaries: "Everyone needs a boss," huffed one benefits director from a large U.S. company. "Medicine in the U.S. has operated on a wide-open basis for too long." And managed care, as the concept had now become known, wasn't popular just with CEOs. Advocates of universal health care were seizing on it, too. If managed care could provide more care for less, they realized, it might be possible to cover everybody without finding new sources of money. Clinton had become one of the believers; his plan envisioned most Americans getting their insurance through private managed care plans. His embrace, in turn, reinforced the business community's determination to switch over their employees.

By the time Clinton's plan failed, there was no turning back: within a few years, nearly every American with private insurance would have some form of managed care. But, along the way, the evolution of managed care had taken a critical turn. When employers had begun clamoring for managed care, few of the idyllic nonprofit group practices that had captured Ellwood's imagination were in a position to meet it. Instead, it was largely the major commercial carriers that had the capital, infrastructure, and business relationships to pick up the new business. By 1997, for-profit companies had two-thirds of the HMO business, up from just over a tenth in 1981.

Early on, some of these for-profit insurers forged alliances with existing group practices, apparently embracing their basic ethos. But the commercial insurers were fundamentally different enterprises, starting with the way they allocated their resources. Every organization involved in the financing of health care has a "medical loss ratio"— literally, the percentage of revenue that ends up being spent on actual patient care, as opposed to overhead, marketing, or profits. (Perhaps tellingly, in the terminology of the insurance industry, the "loss" referred to the portion of premiums spent on actual medical care— not the money going to profits and overhead.) Outfits like Kaiser or the Group Health Cooperative of Puget Sound, a Seattle-based group practice widely recognized for its superb standards of patient care, traditionally had ratios between 85 and 90 percent. At for-profits, the ratio was typically between 70 and 80 percent. If the ratio crept higher, executives could expect a swift rebuke from Wall Street.

But even more important than ownership status, perhaps, was the very structure of the health plans taking over the managed care business—and how they differed from the ones that had first captured the imagination of reformers like Ellwood. Mayo and the old Kaiser plans were true "group practices"—that is, teams of doctors and other health care providers working under one roof (figuratively if not literally). But replicating that model all over the country would have required herding physicians into new organizations while spending heavily on "brick and mortar" to construct new facilities. At a time when insurers were trying aggressively—and quickly—to lure customers by offering lower prices, that didn't seem like an ideal strategy. So the insurers increasingly turned to a different, less structured model of health care delivery and finance: negotiating with doctors already in practice on their own, then constructing networks of "approved" doctors consisting of the ones willing to provide sufficiently deep discounts. Instead of calling themselves HMOs, these entities called themselves preferred provider organizations. In some instances, physicians merged into financial groups that collected monthly fees from members (just like HMOs did) but continued to practice medicine separately (just like doctors in private practice always had). These organizations called themselves independent practice associations, or IPAs.

Most consumers at this point probably couldn't tell one acronym from another. But the distinction was crucial. While the original cooperatives and HMOs had uniform treatment guidelines, these were set largely by the staff physicians themselves, with considerable leeway for special cases. At Group Health of Puget Sound, the medical director summed up the philosophy this way: "If a doctor orders it, it's covered." The new managed care plans set their own criteria and imposed it on their physicians from afar, frequently requiring the physicians to obtain prior approval for referrals, tests, and procedures.

Even more important (and far less understood) was the way the overall financial arrangements in managed care had evolved. Paying doctors a salary, as the old group practices had done, buffered the physicians from incentives to economize excessively; physicians who ordered too many tests might catch some grief at the next performance review, but on a day-to-day basis this didn't have much to do with their income. By contrast, the new plans made much greater use of financial incentives to steer doctors' behavior, basing as much as half an individual physician's income on performance criteria—criteria that, for the most part, considered how well the doctor had cut down on expenses, not how well he or she had treated patients.

If the old face of managed care had been Michael Shadid, its new one was Leonard Abramson, the founder of U.S. Healthcare. Unlike the physicians who first dreamed up managed care, Abramson, a pharmacist by training but an entrepreneur by trade, was openly contemptuous of the medical profession—perhaps because of a long, unpleasant hospitalization he'd endured as a child: "We have created an environment in which providers have become better at billing than they are at practicing their craft," he wrote in 1991, in a manifesto about restructuring the health care system. And while U.S. Healthcare, which Abramson had established in the 1970s, won raves from experts for its sophisticated data systems, it also became notorious for its adversarial relationship with physicians, its aggressive intervention in medical decision making, and its creative use of financial incentives to curb the use of medical services. Among its more famous initiatives was a policy authorizing just one day of hospital care for new mothers who'd had normal deliveries.

These efforts helped U.S. Healthcare to offer prices that undercut its competitors, swelling its enrollment; Abramson's annual pay as CEO reportedly reached $10 million. When he sold the company to Aetna in 1996, he walked away from the deal with a $500 million payout—landing him on the Forbes 500 list of "Richest Men in America." "By applying American capitalism," he proclaimed, "I have realized the American dream."

The Texas Medical Association is among the strongest organizations representing physicians in the entire country. And that power helps explain why Texas physicians were, for most of the twentieth century, among the most successful in keeping managed care at bay. By 1991, though, the allure of more affordable health insurance had simply become too strong. First it was Southland convenience stores, then Texas Instruments, then American Airlines—one by one, the state's largest employers began offering managed care as an option, sometimes the only option; and small businesses and government employers were quick to follow. At Southwest Airlines, the benefits department explained the switch to employees by way of an analogy, distributing posters stating that the company could buy 2.2 million bags of peanuts with the money spent on one day's worth of employee health care. "The education program is trying to get workers to use health care only when it is needed," the airline's director of benefits and compensation explained. "A lot of the costs are unnecessary." By 1994, only one-fourth of all Texans in medium- to large-size companies still had fee-for-service coverage.

The Hilsabecks had entered the world of managed care in 1991—though, like many Americans, they were barely conscious of the switch. They had chosen PruCare, the managed care plan run by Prudential and offered by Steven's bank, largely because it was cheap. Also, Prudential was a well-established, national brand name that they felt they could trust—though, truth be told, they really didn't expect to need

good coverage. After all, they were a young, economically successful couple in excellent health. Although this changed when the twins
were born, the Hilsabecks had only good things to say about PruCare
in the early days. The doctors' bills and hospital bills for the twins'
neonatal care ran into hundreds of thousands of dollars. PruCare covered virtually everything. All that was left for the Hilsabecks was a
$50 co-payment.

That was a pretty big deal—Elizabeth actually called Prudential
to say thanks—because by this point she had her hands full with the
twins. When they first came home, after four months in the hospital,
Sarah seemed to be in more trouble than Parker: she had severe acid
reflux and was slow to gain weight. She also had apnea, a condition in
which breathing stops randomly, which the doctors treated by administering regular doses of caffeine. But within a few more weeks Sarah
stabilized, while Parker seemed headed in the opposite direction. He,
too, had problems with his lungs. Every four hours, Elizabeth had to
give him intensive physical therapy, which consisted of banging him
on the back hard enough to dislodge the mucus that was building up
in his airways. At one point Parker developed a severe respiratory
infection that sent him back to the Austin children's hospital for a few
more days.

After Parker had recovered from the infection, Elizabeth noticed
another problem: he was barely moving. Sarah would wiggle her limbs,
turn her head, and generally thrash about as babies do when they're
learning to control their muscles. Parker would just lie on his back,
crying, his arms fixed in a "W" position. At first the pediatricians told
her not to worry—that boys sometimes take longer to develop than
girls. But in the fall of 1993, when the twins were seven months old,
tests finally showed that something was indeed wrong with Parker: he
had cerebral palsy, a disease that destroys muscle control. And while
nobody could say for sure why he had acquired the disorder—it could
have been because of the premature birth or because of a temporary

loss of oxygen while he was in the hospital for the lung infection—the doctors told Elizabeth they were certain Parker would never walk without physical therapy. Even then, they warned, Parker might be confined to a wheelchair for life.

Parker's physicians recommended a therapy group in downtown Austin. The drive from Lakeway took fifty minutes in each direction; Elizabeth had to make it three times a week, with Sarah coming along for the ride. But at least it didn't cost anything. Although the therapists billed their services at around $150 per session, they, like the doctors, were on PruCare's list of approved providers. And this meant that the Hilsabecks' insurance would cover it.

Or so Elizabeth thought. In early 1994, a few months into Parker's treatment regimen, a clerk from the therapists' office pulled Elizabeth aside: several hundred dollars' worth of Parker's bills remained unpaid. The clerk said the therapists would continue with the treatment, on the assumption that this was just another snafu at the insurance bureaucracy—the type with which the center was already quite familiar. Elizabeth, in turn, promised to make inquiries and get the payments going again. But when Elizabeth called to find out why PruCare had forgotten to pay the bills, she was surprised to hear that the lack of payment wasn't an oversight at all: the company had determined that the services in question weren't eligible for coverage anymore. As Elizabeth recalls, a customer service representative told her that PruCare would cover no more than sixty days of therapy—over the course of a lifetime. Since Parker had exceeded that limit, PruCare had stopped paying.

Elizabeth says that she began to cry, explaining that she and her husband didn't have the money to cover more than $1,000 a month for physical therapy—and that without the therapy, her son would lose whatever chance he had to walk. The PruCare representative was sympathetic, according to Elizabeth, but replied that the family's only recourse was to file a formal appeal with the company—requiring several letters and, quite likely, appearances before a review board at the PruCare regional office, with no guarantee of success. "When was I going to find time for that?" Elizabeth thought, and decided to make two other appeals first—one to an official at the Texas Department of Insurance, and one to the benefits director at Steven's bank. Both proved helpful. It turned out the Texas state employee had a child with cerebral palsy, too. Within a few weeks a letter from Prudential arrived, reversing the earlier denial of benefits. The original interpretation of the Hilsabecks' policy was wrong, the letter stated; the sixty-day limit was per year, not per lifetime. "Please accept our apologies for any inconvenience we have caused you due to this error," the letter stated. "Your concerns will be submitted through our Quality Improvement Process."

Although that episode was over, coverage from PruCare remained problematic over the next few months. In the fall of 1994, the company refused to pay one bill because it claimed the therapist wasn't on its approved list of providers. In fact, this was the same person who had been providing therapy all along. But somewhere between the therapist's office and the PruCare approvals department, somebody handling the bill had written down the wrong identification number for the therapist. It took yet more letters and phone calls from Elizabeth to straighten out the confusion.

By this point, Elizabeth and Steven wanted nothing more to do with PruCare. Among other things, they were paying for all the therapy sessions after sixty days—at a cost of several thousand dollars beyond the approximately $5,000 a year they were already spending on premiums. They quickly depleted their savings, covering the additional ongoing expenses by having Elizabeth take a part-time job at night.

Getting different health insurance was no simple matter, though. As it happened, Steven was entertaining a job offer from another bank, one that would have paid him a much higher salary. But after the disagreements with PruCare, the Hilsabecks knew enough to check about the insurance coverage first. Elizabeth still remembers calling the HMO that would have provided benefits at the new job, detailing Parker's condition and his treatment needs. According to Elizabeth, that prompted a customer service representative to ask, "When is he getting over the cerebral palsy?" When Elizabeth explained that he would never "get over" it, that it was a permanent condition, and that the physical therapy was to build up his muscle strength so he could have some independent physical movement, the representative responded that the company would not cover the therapy at all. The policy, as Elizabeth came to understand it, covered only "rehabilitative" services. And since Parker would be learning to walk for the first time rather than regaining a skill he had lost, the insurance company deemed the therapy "habilitative" and thus beyond the scope of coverage.

Steven kept his job—and the coverage that went with it. And as discouraged as the Hilsabecks were about their travails over insurance, they took strength in Parker's steady medical progress. In the fall of 1994, just before his second birthday, he'd learned to crawl. After the new year, he learned to pull himself up and even take a step while holding onto a table. But the Hilsabecks were left living from month to month—and now the doctors were saying that Parker needed some new medical equipment to help him gain more muscle control. According to Elizabeth, PruCare wouldn't pay for it, arguing that it wasn't "medically necessary," since there was no guarantee that a child with such severe muscle problems would ever walk. Although the company's statement was true, the Hilsabecks didn't want to give up even a modest chance.

They contemplated divorcing, just so that Elizabeth could become a low-income, single mother who would qualify for Medicaid—which would have covered Parker's services more comprehensively than PruCare. But they rejected that notion and reluctantly settled on another course of action: Selling the Lakeway house and moving into a mobile home beyond the outskirts of town. The mobile unit would turn out to be about half the size of the Lakeway split-level, making for a tight squeeze with all of the twins' medical equipment. And there wouldn't be many opportunities for decorating since applying paint or wallpaper would have voided the warranty. ("It felt like living in a paper towel," Elizabeth would later recall.) When the time finally came to sign the sale papers, completing the Lakeway sale, Elizabeth was crying and shaking. All she could think about was the peach-and-green nursery—and the long hours she had spent there, on the rocking chair, singing her babies to sleep: "Sweet dreams 'til sunbeams find you, sweet dreams and keep your worries behind you . . ."

———

Garbled communications with insurance companies' representatives. Confusion over the scope of benefits. Difficulty getting approval for ongoing treatment. These were all hallmarks of contemporary managed care, experiences even families with relatively routine health care needs had encountered as they tried to navigate the new world in which insurers insisted on asking questions—often many questions—before agreeing to pay. But the Hilsabecks were in many respects fortunate. Other stories of apparently excessive zeal in managed care ended with even more serious consequences. There was the case of Nelene Fox, a woman in California with breast cancer, who died after her HMO denied coverage for a bone marrow transplant. There was the story of the Adams family in Georgia, whose infant son went into cardiac arrest on the way to the emergency room, because the family's HMO insisted on sending him to a hospital forty-five miles away. Although the boy lived, he suffered permanent disabilities. And then there was the story of Bryan Jones, an infant in New York City who went home from the hospital with his mother just twenty-four hours after birth, allegedly because of his HMO's coverage directives. He died a few weeks later, because a heart defect—one typically diagnosed shortly after birth—went undetected until it was too late.

A few of these cases led to successful lawsuits against the insurance companies: separate juries awarded $45.5 million to the Adamses and $89.1 million to the family of Nelene Fox, though in both cases subsequent out-of-court settlements reduced the amounts. But even those stories that didn't end up in court resonated with a public already wary of managed care, because it was interfering with their traditionally unfettered access to doctors, hospitals, and treatments. Public frustration expressed itself most famously in 1997, when the actress Helen Hunt's expletive-laden reference to HMOs in the movie *As Good As It Gets* began eliciting spontaneous applause from audiences around the country. "It got more response than anything else in the movie," wrote a columnist at *Newsday*. "I was clapping, too. All of us, by now, have learned to hate an HMO."

The managed care industry bristled at such statements, insisting that the data showed that their patients fared just as well as patients in old-fashioned fee-for-service plans. And the industry had a point. Scholars could find no substantial evidence that the American population, in the aggregate, was suffering worse health outcomes because managed care had spread. Studies also showed that, true to their original spirit, HMOs covered more basic preventive services. According to one study by the Kaiser Family Foundation, only 81 percent of traditional fee-for-service plans covered annual obstetrical exams for women and only 63 percent covered regular adult physicals. Ninety-seven percent of HMOs covered both—in no small part because, given their greater ability to control costs, they could afford to offer a wider range of services while maintaining competitive prices.

The managed care industry and its allies also protested that the "horror stories" were a lot more complicated than they appeared at first—which certainly seemed to be the case. The tale about Bryan

Jones, the infant in New York City, had appeared on the front page of the *New York Post*, featuring pictures of the family alongside one of the newspaper's typically understated headlines: "What They Didn't Know about HMOs May Have KILLED THIS BABY." And, sure enough, it was the infamous U.S. Healthcare that had apparently rushed him and his mother out of the hospital after just one day. But it was the doctors and nurses who examined Jones in the first week who failed to diagnose the heart deformity that eventually killed him. Yes, they were on U.S. Healthcare's list of preferred providers. But to what extent was U.S. Healthcare responsible for their providers' behavior? And would a longer hospital stay have made a difference?

In other cases, the treatment an HMO refused to cover was very expensive and of questionable value to the patient. The story of Nelene Fox, the woman in California with breast cancer, was an example. The bone marrow transplant became a highly sought treatment for breast cancer in the 1980s, not long before Fox was diagnosed, because it had been shown to work for other cancers. But the few studies suggesting that it might work for breast cancer were hardly clear-cut, since they did not compare survival rates of people who had the procedure with rates of those who hadn't. (It was hard to find people for such studies because once doctors started talking about the treatment, nobody wanted to be part of the control group that went without it.) Not until 1999, years after Fox's death, did more definitive research come out. And those studies suggested that a bone marrow transplant would help only a tiny fraction of patients—if, indeed, it helped anybody at all.

Might Nelene Fox have been one of those few people whom the procedure might help? How much was it worth spending to find out? Those were complicated questions, as much philosophical as medical, about which serious people could—and did—disagree. And they were common in every country in the world where the skyrocketing cost of medical care was forcing citizens to contemplate the true value of new or experimental treatments. But because other countries had universal health insurance systems, administered in one way or another by the government, the decision makers were accountable to the voters. That had been one goal of Clinton's health care plan, too.

In the absence of Clinton's proposed rules or any other strict government oversight of managed care, the ultimate authority for decisions about coverage remained the insurance industry itself. Each insurer had its own rules. And while the rules were typically written up by medical directors on the staff, they tended to conform closely to the guidelines of a consulting firm called Milliman and Roberts, which published annual guidelines for appropriate medical care. The process through which Milliman came by its guidelines was something of a black box: Milliman refused to reveal many details, calling it proprietary information. The best hint about the process was a list of physician advisers at the beginning of each Milliman guidebook. Milliman said those lists were proof that its guidelines were medically sound, but not everybody agreed. When a pediatrician in Houston discovered his name in a Milliman guidebook, he protested and eventually sued, angry in part because he thought the recommended pediatric guidelines were clinically unsound. It turned out that he had ended up on the list because his hospital's administrators had agreed to participate in discussions about it. (The suit was eventually settled out of court. The terms were confidential—except for the fact that Milliman wrote the physician and a colleague letters of apology.)

And the pediatric guidelines weren't an isolated instance. If the old hands-off approach to paying medical claims favored overtreatment, the new one seemed seriously biased toward undertreatment. Plans offered financial bonuses to doctors for ordering fewer tests, performing fewer procedures, and prescribing fewer drugs, while doctors who consistently ran up higher costs occasionally found themselves "delisted"—insurance parlance for a provider whose contract an insurance company has terminated. (As an article in the magazine

Medical Economics explained in 2004, while managed care firms frequently boast about selecting doctors based on quality, "cost remains king.") One study in the *Archives of Internal Medicine* found that in 1981, the average patient hospitalized for a broken hip left the hospital after twenty days; in 1998, the average patient left after just seven days. Although better surgical techniques and the expanding availability of rehabilitation hospitals had something to do with this, the authors of the study—physicians at Mount Sinai Hospital in New York—concluded that the patients were also in relatively worse health as they were discharged.

This was quite a contrast to the way the old group practices operated. Plans like Group Health in Seattle had embraced what the journalist Robert Kuttner called a "social ethos." But rather than reward such conduct, the market actually seemed to punish it. Perhaps the best example of this was the evolution of Kaiser Permanente, once considered a model managed care organization. Kaiser's domination of the California insurance market gave it the financial power other old-fashioned HMOs lacked; during the 1990s, it had the resources for the same kind of explosive expansion as the for-profits. But the more it competed with the for-profits, the more it seemed to act like one. Stories spread that Kaiser was implementing crude, ill-conceived cost-control efforts. (It was a Kaiser plan that had sent that sick boy in Georgia to the faraway emergency room, nearly killing him, too.) In 2006, an investigation by the *Los Angeles Times* revealed that Kaiser had refused to authorize kidney transplants for some of its patients in the San Francisco area, even if the refusal meant turning down available organs, because the costly procedures would have taken place at non-Kaiser facilities. Kaiser insisted that the stories were overblown (although it apologized for refusing the transplants and subsequently abandoned its program), but critics saw in anecdotes like these yet more proof that competition was squandering whatever promise managed care had.

The idea that modern managed care was sacrificing medical benefits for the sake of financial ones was particularly distressing to Paul Ellwood himself. Although he thought the horror stories included considerable media hype, and although he thought income-sensitive physicians were a much more pernicious influence than for-profit insurers, he too perceived market dynamics diverging from his original vision. "The idea was to have health-care organizations compete on price and quality," Ellwood eventually said. "The form it took, driven by employers, is competition on price alone."

As proof, Ellwood spoke frequently about a Jackson Hole collaborator, one of the original HMO idealists, who lived in fear that his plan's generous benefits for AIDS would become well publicized, thereby attracting all the AIDS patients and destroying the plan's delicate actuarial balance. But Ellwood also had his own personal perspective on this shift, as he explained in an article in the *New York Times Magazine* in 1996. Ellwood, then seventy, had recently had surgery for a hernia. On the car ride home, he experienced the kind of common postoperative side effects—nausea, dizziness, and vomiting—that in past generations might have kept him in the hospital longer, even overnight. When Ellwood saw the surgeon again, he related the episode. "Ellwood," the surgeon said, "it's . . . your . . . own . . . damn . . . fault."

While it was the dramatic controversies over potentially lifesaving treatments that frequently grabbed headlines, it was the more mundane, everyday hassles of managed care that sometimes seemed to generate the most ill will. In 2000 and 2001, several groups of or representing physicians, including the Texas Medical Association, joined in a class action lawsuit against the nation's top insurers, claiming that they had rigged their authorization and payment systems in order to

deny care unnecessarily and delay, if not withhold altogether, appropriate payment for services.

Although the insurers denied that any such effort was under way,
Elizabeth Hilsabeck felt that this was exactly what had been happening to her family. And it seemed to be happening again, in April 1995,
when another dispute over Parker's physical therapy erupted. It was
the same old issue, the sixty-day limit, but with a novel twist. This
time PruCare was insisting that the sixty days be consecutive. The
Hilsabecks were dumbfounded. Was Parker supposed to get therapy
every single day for two months, then stop altogether?

What really upset them was that PruCare hadn't said a word
about the change, which had apparently taken effect at the beginning of the year, until Parker had gone past the sixty days—in other
words, after the bank's open enrollment period had ended. (Open
enrollment is the period when employees can switch insurance plans,
if their employer offers more than one option.) They complained to
the bank's benefit manager, but it turned out that she didn't know
about the policy change, either. The lone communication about
benefit changes she'd gotten from PruCare was in February. And
that letter notified beneficiaries about a change in the coverage of
transplants. It said nothing about physical therapy, although she felt
that the omission was almost beside the point, since this letter had
also arrived after open enrollment: "I can tell you that had I elected
to go into PruCare effective January 1, 1995, and was facing the
above transplant and was not told until February 14, 1995, that the
procedure was not covered, I would be very upset," she wrote in her
own letter to PruCare. "If changes are made to the plan provisions,
companies should be notified prior to the open enrollment period
so that employees can make elections of coverage based on plan
provisions."

PruCare eventually backed down on this requirement, too, citing
its failure to inform beneficiaries about the shift: "We certainly did

have some problems with notification. . . . I'm not going to dispute that," a vice president of PruCare said. But precisely because the company hadn't repudiated the policy itself, Elizabeth worried about what would happen the following year. And she was already busy doing something about it.

In the midst of the various billing controversies, Elizabeth had started contacting attorneys, support organizations, and advocacy groups for advice. That's when she discovered that the entire managed care system operated in a sort of legal no-man's-land: ERISA—the same law that encouraged large companies to finance their own benefits, thereby disrupting the risk pooling of insurance—also gave most managed care plans what amounted to nearly a blanket immunity from state health insurance regulations. If a consumer believed an HMO had denied coverage of "medically necessary" services, which is precisely what Elizabeth thought PruCare had been doing to Parker, the consumer had no legal recourse except to sue for the cost of the treatments itself, an amount few consumers felt justified the effort and expense of a lawsuit, particularly given the lengthy legal process.

Eventually Elizabeth got the idea to form an advocacy group, in order to raise awareness of how managed care affected children with disabilities. She called it Texas Advocates for Special-Needs Kids (TASK), and she asked some of the health care professionals who'd helped Parker to serve on the board. As for the membership, she went about that a little more deviously—by snatching a list of patients from a business desk at the therapists' office, then sending out a mass mailing explaining her own story and asking other parents to write in with theirs. The response overwhelmed her. About 500 families ultimately ended up joining, in many cases after parents called Elizabeth at home, sometimes in the middle of the night, for hours-long conversations exchanging tales of hassles over managed care. At the end of one such call, about a little girl in El Paso suffering from a condition even more serious than Parker's (Elizabeth can't remember what anymore), the mother asked what Elizabeth could do to help. "Well," said Elizabeth, "I'm living in Austin now, I am not employed, and I am one pissed-off mom."

Of course, Elizabeth had some help, too. Two of the groups she had originally contacted—the Texas Physical Therapy Association and the Texas Medical Association—had been trying to pass a "patient protection" law that, among other things, would have ended the insurance industry's exemption from legal accountability. But those efforts had stalled, in no small part because the groups seemed more interested in protecting their own turf, and incomes, than pursuing the public good. Organized medicine and its allies needed a more appealing public image—something that Elizabeth and Parker, with his blond hair and blue eyes, supplied perfectly.

Elizabeth accepted the offers of assistance, aware of her political utility to the professional groups but willing to play along because she had come to believe that reforming managed care would tame HMOs. Over the ensuing months, the cause became something of an obsession for Elizabeth, the outlet for all her frustrations. She started wheeling Parker down the halls of the state house, paying unscheduled visits to legislators who didn't support the HMO reform bill, and soon enough stories about kids like Parker started filling the pages of newspapers in Texas.

Elizabeth's lobbying helped fuel the growing public resentment of managed care and, eventually, that resentment overcame the staunch resistance of insurance and business lobbies—which argued that the regulations would simply pad the pockets of doctors and drive up the cost of insurance, without actually improving medical care. Later in 1995, the legislature finally passed a measure to reform managed care—only to have the governor, George W. Bush, veto it. But the advocates for the measure, both inside and outside the legislature, continued the push, holding a series of statewide hearings where parents like Elizabeth testified for the record about their experiences. At the next session of the state legislature two years later (the Texas legislature meets only every other year), another nearly identical bill passed by an even larger margin. Bush decided to make peace with it rather than continue a political fight he was losing. He let the measure pass into law without his signature.

Texas thus became the first state to pass a comprehensive managed care reform bill that allowed beneficiaries the right to sue HMOs over denials of treatment. Nationally, those sympathetic to the cause hoped—while those hostile to it feared—that the victory would be a harbinger of similar actions in other states and, eventually, in Washington, D.C. But both predictions proved exaggerated. Reform of managed care did become a central health care issue in the congressional elections. It also came up in the 2000 presidential race when, to the dismay of Hilsabeck and many others in Texas, Bush took credit for the HMO law that he had fought until almost the very end. (Despite this, Elizabeth, a lifelong Republican, voted for Bush anyway.) Bush also promised to enact a similar measure if he became president. But once he got to the White House, he indicated he would not sign a bill that didn't remove some of the protections—chief among them the right to sue—that advocates believed were essential. Within a few months, anxiety over the affordability of health insurance displaced the managed care backlash on the health care policy agenda. (And, in the meantime, 9/11 had nudged health care itself off the political agenda.) That was the last anybody heard of the measure.

Back in Texas, some health professionals said they noticed a change in insurers' attitudes, albeit a small one—possibly suggesting that the new rules were having some effect. But for reformers, even that victory proved fleeting. In 2004, the U.S. Supreme Court invalidated most of the Texas HMO law, because the federal law giving managed care plans their special legal status (the law Congress could have changed but didn't) superseded it.

Managed care by then had already adapted on its own, in both Texas and the rest of the country. By the late 1990s, insurance companies had learned how to soothe a rebellious public, by ameliorating the features that most rankled average beneficiaries, while leaving in place some of the hidden incentives and limits that had made trouble for families like the Hilsabecks. What's more, as managed care loosened its restrictions it sought to recover the money in other ways. Chief among them: Targeting companies with generally healthy employees (or, within a company, the most healthy groups of workers), as insurance companies had been doing ever since commercial carriers got into the business during the 1940s and 1950s.

Managed care had frightened middle-class Americans because the closed networks of doctors and clunky preapproval requirements seemed so different from the kind of coverage to which they were accustomed. But ultimately it was the similarities between managed care and old-fashioned commercial insurance—in particular, their shared propensity for marginalizing those people with the worst medical conditions—that probably caused the most actual hardship. In that sense, managed care hadn't so much altered the evolution of American health insurance as reinforced it, moving toward a system that left the most medically vulnerable at even greater risk than before.

———

As for Elizabeth herself, her political crusade and the difficulties that provoked it had taken their toll. A year after losing her home in Lakeway, she lost her marriage. Like any divorce, this one had many causes. Studies have found that couples with children who have severe disabilities are more likely to divorce, apparently because even relatively strong marriages have a hard time withstanding such stress. But

Elizabeth says she thinks she and Steven might have made it if, on top of all the other strains that the twins' medical problems had caused, they didn't also have to fight with their insurance company and deal with the threat of financial loss. (Steven, for his part, agreed that the insurance-related stress strained their marriage.) Like Gary Rotzler and Janice Ramsey, Elizabeth had been left to face the financial burden of medical care—or part of it, anyway—on her own. And, like those other two, she'd paid a price for it.

Still, reflecting on the whole experience in Austin one day years later, Elizabeth said she had no regrets about what she gave up and no anger over the fate life handed her. After all, she figured, the struggles had produced one unambiguously positive outcome: Parker had finally learned to walk. "When people tell me that they're really sorry that I have a child with cerebral palsy, I say, 'Don't be.' God gave him to me because I'm that bitchy and that stubborn and I don't give up very easily."

FOUR

SIOUX FALLS

Morning shift inside the pork by-products room at J.P. Morrell would begin when the hogs, freshly slaughtered and dangling by their hind legs, rolled in on a metal chain suspended from the ceiling. The first men to attack the carcasses would use chain saws, slicing open the pigs' fattened bellies. The next group would wield carving knives—some hacking off the internal organs, which would be ground into sausage; others removing ears, lips, and other parts that could be sold as delicacies. Eighty men could process up to 1,000 hogs an hour this way, removing from the carcasses everything except bone and the choice cuts that would become fresh pork in another part of the plant. But before the hogs could get that far, there was one last body part to claim: what was left of the pig's head. And that's where Lester Sampson's job came in. After a coworker had cleaved the cranium clean from the torso, Lester would place it inside a large metal vise. Reaching high to grasp a handle, Lester would pull down hard, cracking open the skull. Then he'd place the crumpled mass on the table in front of him and scoop:

brains into one pan, everything else into another. Every few minutes runners would come by and replace the full pans with empty ones, so that Lester could repeat the process—reach, crack, scoop—more than 100 times a day.

It was not an exciting job, and the skull contents didn't smell very good. But it paid sixty cents an hour, which was more than adequate by the standards of 1942. Lester, still just seventeen years old, was technically a year too young for a job at Morrell, the city's largest employer. But he had schemed his way onto the payroll by lying about his age—a deception the company was probably happy to indulge, given the domestic labor shortages during World War II. And although the draft would make that first stint at Morrell a brief one—Lester ended up serving more than five years in the navy—he returned many years later, after he and his new wife, Audrey, learned that they were expecting their first child. Lester would miss the romantic elements of navy life, particularly the exotic ports of call like "Rio Janero," as he called it; but he figured that Morrell would provide a lifetime of good pay and benefits, just as it had done for his father and literally thousands of other workers who had become part of the Morrell family since its establishment in Sioux Falls, South Dakota, four decades before.

Lester still felt that way in 1985, when, having put in the thirty years necessary to qualify for the company pension plan, he prepared to leave Morrell. By then, he'd worked in virtually every department. And even the more unpleasant assignments—like working in the dank hide cellar, where employees lugged sixty-pound beef and sheep pelts on their backs—seemed worthwhile, given the lifestyle he'd been able to have. Together with money Audrey had made after returning to work, as a box sorter for United Parcel Service, the couple had saved enough to buy a mobile home on Lake Madison, a medium-size fishing pond about forty-five minutes to the north. On weekends and vacations during the warm weather, Audrey would tend to the garden while Lester sat on the back porch with a line in the water, getting "all goofy" as he pulled up walleyes, perch, and blue gills. Eventually the Sampsons would sell their modest Sioux Falls home altogether, using the money it fetched to buy a Ford extended cab pickup and a trailer camper with matching brown trim. In the winters, when Lake Madison became effectively uninhabitable, the Sampsons would drive the truck and the fifth wheel to Arizona, where they'd soak up the desert heat at local campgrounds.

It was the kind of retirement to which the Sampsons had long aspired. And, as far as they could tell, only one problem threatened to undermine it: Lester's health. A former smoker, Lester had a history of serious cardiac and respiratory problems, including a heart attack when he was just fifty-seven years old and still working at Morrell. But although Lester realized that medical care could only stop his decline, not reverse it, he and Audrey comforted themselves with the knowledge that financing his treatment, which was bound to get more expensive over time, would not be a concern.

They knew this because Lester had two reliable sources of health coverage. One was the federal government, which paid for basic hospital and physician services through Medicare. The other was Morrell, which offered retirees a lifetime of health benefits. Those retirees under sixty-five, and not yet eligible for Medicare, were entitled to exactly the same coverage as current workers got. Those over sixty-five, like Lester, were entitled to supplemental insurance that covered whatever Medicare didn't, such as out-of-pocket payments for hospital stays and prescription drugs. Morrell's coverage was extremely generous; it even extended to spouses and widows. Employees who wanted it had to relinquish part of their pensions, but that was a deal workers like Lester gladly made, since they understood the potential financial devastation that came with serious illness. Guaranteed health coverage was worth more to them than extra money. They craved security more than luxury.

But the security Morrell provided turned out to be fleeting—as Sampson and about 3,000 of his former colleagues discovered one day in 1995, when they received letters notifying them that the company would be stopping its retirement health benefits. Former workers still under sixty-five would be allowed to purchase Morrell's coverage on their own at full price, but only for a year, at which point they'd have to find alternative policies. Older retirees had just three months to find new supplemental coverage. If they couldn't get it, they'd have to go on Medicare alone.

Morrell offered a straightforward rationale for its actions: it said it could not afford to subsidize such generous insurance for nonworkers because it had to compete with companies that were stingier with benefits or had fewer retirees to support. "Morrell has been in a struggle to survive for the last four or five years," explained the company president. And while that hardly satisfied Lester and his former co-workers, who believed the company had broken a promise, Morrell's action was hardly unusual for the time. In the 1960s and 1970s, it had become customary for large companies, particularly large manufacturing firms, to give their retired workers health insurance. By 1988, more than two-thirds of large employers were promising to provide their current workers with some form of supplemental health insurance during retirement. But by 2004, the proportion had dropped to just over one-third, with such industrial icons as Bethlehem Steel and General Motors among those reducing or eliminating health insurance for retirees—provoking similar dismay among their former workers.

Medicare certainly softened the blow for these retirees. But Medicare, a program frequently criticized for its supposedly excessive generosity, has always required the elderly to pay for much of their health care out of their own pockets. Part A, the hospital portion of Medicare, has cost sharing and deductibles. Part B, the portion that covers physicians' services and some outpatient services, has premiums. And there are many health care expenses that Medicare covers inadequately or not at all, such as prescription drugs and nursing home care. Since the 1970s, these out-of-pocket costs have risen faster than seniors' income—which is another way of saying that retirees' bills for health care have risen faster than their ability to pay, even with Medicare kicking in its share.

More affluent seniors have been able to absorb these costs without too much difficulty, either by dipping into their own savings or by purchasing individual Medigap policies that cover what Medicare doesn't. The least well-off seniors, meanwhile, have been able to get assistance directly from the government, through Medicaid. But the seniors in between, like the ones who worked thirty years on the line at Morrell, had depended on company-provided benefits to make up the difference between what Medicare covered and what they needed. When those company benefits vanished, they found themselves in the same situation as Lester and Audrey Sampson, making painful financial trade-offs about the way they would live out the remainder of their lives—the kind of trade-offs that they had spent their working years hoping, and laboring, to avoid.

———

Up through the late 1940s, the debate over how to pay for medical care in this country really didn't treat retirees as a separate group. But that changed after Truman's failed bid to establish national health insurance, when politics—and the evolution of private health insurance—forced the "special problem of the elderly" onto center stage.

By that time, enrollment in private insurance had expanded to most of the working-age population, primarily through job-based coverage. Indeed, widespread satisfaction with those arrangements had been a major reason Truman's bid failed. But the elderly were not sharing in this progress, at least not fully. Most commercial insurers wouldn't allow beneficiaries to keep coverage when they retired from their jobs, particularly if they were past the age of sixty-five. Nor would they sell seniors insurance directly, since the elderly—almost by definition—were precisely the kind of high medical risks insurers tried to avoid. As a result, just 40 percent of seniors had health insurance by 1957. The figures improved in the next few years, as the insurance industry, in an apparent effort to prove that government intervention was unnecessary, aggressively offered coverage to more seniors. But even though two-thirds of the elderly would eventually get some form of insurance, the coverage was frequently thin, with exclusions or high cost sharing for any preexisting coverage. Overall, by 1961 insurance was covering about 7 percent of elderly people's total medical bills.

In Washington, the frustrated proponents of national health insurance and health care reform hadn't given up. But they had made a strategic decision to move toward their goal incrementally. And the elderly were an obvious place to start. As the political scientist and historian of Medicare Theodore Marmor has noted, "The aged could be presumed to be both needy and deserving because, through no fault of their own, they had lower earning capacity and higher medical expenses than any other age group." Reformers bided their time during President Dwight Eisenhower's terms, using Congress to hold hearings publicizing the plight of the elderly. In the 1960s, political momentum finally seemed to swing the reformers' way when John Kennedy promised guaranteed health coverage for the elderly—"Medicare," he called it—as a central tenet of his domestic agenda. Lyndon Johnson's succession to the presidency after Kennedy's assassination, along with the large Democratic congressional majority he helped to construct in the 1964 election, made passage of a program seem even more likely.

But the architects of the Medicare proposal, many of them veterans of Truman's failed campaign, feared that powerful interest groups opposed to government intervention in health care might yet defeat the measure. So they set out to co-opt such groups. To quiet the American Medical Association, Medicare Part B would pay doctors their "usual and customary fee[s]" as long as they were "reasonable." (Initially, there was no Part B—the program wasn't even going to cover physicians' services at all.) To ease the hospitals' worries, Medicare would deputize independent, private intermediaries to handle the actual administration of payments—a job only Blue Cross, which remained close to the hospital industry, was equipped to perform. In addition, Medicare would cover just key services like hospitalization, and only within certain limits. (Hospital coverage would be restricted to sixty days per episode.) This guaranteed a market for private supplemental insurance, while freeing the insurers from any pressure to provide basic coverage for the elderly with serious medical conditions—a task that, internally, many had long since decided they could not do profitably. Although the doctors never backed down, the insurers never fought hard. The hospitals, as desperate to alleviate their charity care as they had been in the 1930s, actually embraced the proposal—giving it enough support that it became law in 1965.

Johnson signed the law in 1965, with an aging Harry Truman at his side, and it was an instant hit with beneficiaries. Soon there was a huge increase in the use of medical services—as well as a substantial decrease in poverty among the elderly (although that also reflected increased Social Security payments enacted at the same time). Medicare also poured money into the health care sector, fueling a boom in hospital construction and research. But as the era of liberalism that had stretched from Franklin Roosevelt to Lyndon Johnson was ending, Medicare became an object of controversy because of its price tag, which was quickly exceeding projections. Partly, this reflected the success of the program in bringing seniors into the mainstream of American medicine. But partly it reflected the fact that concessions to the industry had given doctors and hospitals enormous freedom to raise fees—even beyond levels that seemed "reasonable."

Eventually the cost of Medicare became so big that even some of Washington's most famously antiregulatory conservatives agreed to have the federal government become much more aggressive about dictating payments to doctors and hospitals. In the 1980s, the Reagan administration and the Democratic Congress quietly agreed to introduce a new hospital payment system based on diagnosis, not on services rendered, in an effort to force hospitals to adopt more cost-conscious treatments. Uniform fees for physicians followed shortly thereafter, thereby fulfilling the fears of the old hands at AMA who swore the government would turn to price-setting sooner or later.

These reforms succeeded in their primary goal: containing the cost of Medicare. Although the program continued to get more expensive, thanks to a gradually aging population and the spread of expensive medical technology (which, of course, kept seniors alive longer, thus perpetuating the spending cycle), by the late 1990s Medicare had shown that it could hold down medical costs just as well as private insurance could—maybe even better. But the emphasis on cost control had another impact on seniors: it fostered what the political scientist Jonathan Oberlander has called a "negative consensus" in Washington that stopped the natural evolution of Medicare.

In the 1970s, the standard private insurance policies for working-age Americans expanded to cover new services, such as prescription drugs and outpatient services, that were becoming more essential, particularly for the treatment of the chronic diseases that seniors were likely to have. But Medicare's benefits didn't keep up, exposing seniors to more out-of-pocket expenses. In fact, with the exception of adding dialysis coverage, which was something of a political fluke, serious expansion occurred only once—and only in a botched way that virtually guaranteed its failure.

That instance came in 1988, when Congress passed and Reagan signed a law establishing Medicare's first "catastrophic benefit" package. The law extended hospital coverage past sixty days and limited total out-of-pocket expenses for seniors, all while subsidizing a modest portion of prescription drugs and other expenses. The gaps the law filled were real; this is one reason the program passed with such large bipartisan support. But very few seniors ever reached Medicare's catastrophic hospital limit anyway, so few found this particular service—by far the new law's most important and expensive feature—valuable. Meanwhile, because the government had decided that seniors would have to pay for the new program on their own, with no direct subsidy from the government, it had a substantial premium that was based on income. Had the program been voluntary, it's likely that relatively few people would have enrolled. Instead, it was mandatory—and, ultimately, quite unpopular. In a scene that was one part *Nightline* and one part *Saturday Night Live,* a mob of angry seniors in Chicago chased a frazzled Dan Rostenkowski, a veteran Democratic congressman, into his car. A year later, Congress took the law off the books.

Had the "catastrophic benefit" package focused more on something that seniors really needed—say, by offering something closer to full coverage for prescription drugs, which working Americans typically had by that time—it might have been more popular. Or maybe not. On paper, the "catastrophic benefit" made sense. But at that point many, perhaps even most, seniors were confident that they had enough insurance to cover their needs, and their confidence reflected the fact that many of them had private supplemental insurance already. (In addition, large numbers of seniors were persuaded by a misleading public relations campaign that they would end up paying more for the benefit than they actually would have.) For these people, the grand bargain of Medicare—in which government provided only the basics and private insurance supplied the rest—was working just fine. So why bother having the government do more?

When Lester Sampson's father was working at Morrell, in the 1920s, he had made just $7 a week—a tiny fraction of what his son would earn just two decades later. Lester's father frequently put in ten hours a day, occasionally more, particularly during the winter months. That's when Morrell workers would take ice from the Big Sioux River, which runs past the plant, and pack down the meat cars bound for Chicago. (This was before the era of electric refrigeration.) According to Lester, although the overtime work was technically voluntary, management didn't work very hard to maintain that illusion: "After supper we're going to pull up ice," the bosses would say, "and if you don't show up, don't come back, you won't have a job in the morning."

That environment began to change in the 1930s, when the social unrest of the Great Depression and the labor-friendly legislation of the New Deal emboldened unions to demand more rights and better pay for their members. In Sioux Falls, a local from the Amalgamated Meat Cutters and Butcher Workers of North America represented Morrell's employees. And after the company dismissed twenty-nine union members for staging a three-day work stoppage, the union called a full strike—one that would last two years and would be marked by violent clashes pitting strikers against replacement workers and the police called in to protect them. "The air was filled with brute screams of rage," a correspondent from the local *Argus Leader* reported after one melee. "The picket line gave in the middle. Victorious clubmen, cursing, waving their arms, pounding skulls with their sticks, surged through. It looked like a football game."

In the eventual settlement, Morrell's workers won better pay and new rights, including the seniority privileges that had been an issue in the initial walkout. And although the striking workers did not return immediately—the company kept the replacement workers, agreeing to hire back strikers as jobs opened—the four decades that followed were a time of relative labor harmony, when strikes were infrequent, short, and nothing like the bloody tangles of the 1930s. To hear Lester and some of his colleagues tell it, management and workers had an almost blissful relationship back then, more as members of a common enterprise than as adversaries sparring across the bargaining table and, occasionally, across picket lines. "No man works for me, every man works with me"—that's how one beloved CEO spoke of his employees. And Lester had come to believe such rhetoric. Once, after a serious illness, his supervisors temporarily shifted him over to the "koshering" room, where workers simply sorted cows and sheep as a rabbi decided whether the animals passed muster. When that job was done, after just a few hours, the company would tell Lester to go home and recover, while still giving him a full day's pay. "It was sort of a rehab," Lester later remembered.

The essential ingredient for this relationship was the nation's sustained postwar prosperity, in which Morrell and, ultimately, its workers shared. By the 1960s, after Lester had returned to the plant, starting workers were making $10.90 an hour—a good bit higher than his salary in 1942, even after adjustments for inflation. Overtime paid even more, so that when opportunities came up, workers jumped at them. "Down in the curing room, we'd go after supper. Man, that was the gravy. You could make double there." None of this made a job at Morrell glamorous. But for people in Sioux Falls without a college degree, it was the surest path to prosperity, complete with home ownership and other accoutrements of middle-class life.

And salary wasn't the only financial reward of working at Morrell. Sometime in the 1940s or 1950s, the company had also begun giving its employees group hospital insurance that quickly evolved into virtually seamless medical coverage, including everything from doctors to the drugs they prescribed. A worker at Morrell could pretty much see any specialist, visit any hospital, get any procedure, and pay for any prescription without anything more than a tiny co-payment of $1 or $2. And, as in other heavily unionized industries, these benefits were eventually extended to retirees.

But the business of meatpacking turned out to be no different than forging steel, making cars, or assembling electronic equipment: by the

1970s, Morrell found itself struggling to keep up with the competition. In Morrell's case, though, the competition came not from abroad but from home—in the form of nonunion companies. The freedom from contracts with large industrial unions gave these competitors more flexibility to acquire new technologies, reconfigure their operations to meet changing markets, and, above all, compensate their workers less generously. By the 1980s, for example, one nonunion meatpacker based in Nebraska was paying its production workers $6.50 per hour, $4 per hour less than the salaries at Morrell.

Morrell had quietly undergone a second transition, also common by this time: the British family that had owned Morrell since its founding had sold it to a food conglomerate, based in Cincinnati, that also owned Chiquita bananas. Morrell's British owners had seen themselves as stewards, if not of their workers then at least of an enterprise that benefited from a loyal, long-term workforce. The new owners focused more relentlessly on the bottom line. And it wasn't long before Morrell's new management, eyeing the growing gap between wages at Morrell and wages at its competitors, began pressing for pay cuts.

At first, the union in Sioux Falls grudgingly went along. But in 1989, in the midst of a sympathy strike for workers in another Morrell plant, it decided to make a stand by refusing the company's demand that workers accept an additional $2 reduction in their hourly wage. The company called in replacement workers and, for the first time since the 1930s, violence erupted on the picket lines. While the local reportedly drew $20 million from the international union's strike fund, United Brands, as the parent company was then called, continued to post profits. Workers finally returned to the job in 1991, accepting a wage cut only slightly less severe than the one they had rejected earlier in the strike—leaving many to wonder whether the union had foolishly sought a confrontation it could not possibly win.

The one place where the union had held the line was health insurance. In fact, the one concession the union won from the company in

1991 was a reduction in the workers' contribution to health insurance premiums. And while current workers were gradually being asked for some relatively minimal co-payments for medical services, retirees were exempt even from those changes: they got to keep the same coverage they'd had all along. But the health coverage in general—and retirees' insurance in particular—was becoming as much of an albatross as the old pay scale. By some estimates, the price of all benefits, for present and retired workers, nearly doubled Morrell's labor costs.

At one time, companies like Morrell could afford to ignore such long-term liabilities. But that was no longer possible, thanks not only to the newly harsh business climate but also to a change in the federal rules governing company benefits. Previously, corporate financial reports routinely underestimated the future costs of retirees' health benefits—an illusion that pleased the companies, since it made the bottom line look better; and frequently pleased the employees too, since it encouraged the companies to be more generous with benefits. But in 1990, following a series of scandals involving pensions, the Financial Accounting Standards Board announced that it would begin requiring corporations to make more realistic assessments of future liabilities. Investors quickly soured on the companies that owed the most; the companies reacted by slashing the benefits or getting rid of them altogether.

Morrell was one of those. After reaching a new collective bargaining agreement with the union in 1991, the company announced that it intended to reduce retiree health benefits unilaterally; and since it knew the move would be controversial, it quickly sought a court order recognizing its right to do so. Retirees like the Sampsons were incensed, particularly since they'd given up higher pensions in exchange for what they believed was a guarantee of lifetime health insurance. And although Morrell officials said during the subsequent trial that they had not explicitly promised that the coverage of benefits would last indefinitely, several retirees testified they had been led to believe precisely that. Typical was the recollection of one worker who said company officials had told him that "The only way you can lose that health insurance is if the company went belly up and went broke."

In the end, though, the contents of those discussions turned out to be irrelevant, at least in the eyes of the courts. A three-judge panel broke down along partisan lines, with the two Republican appointees arguing that Morrell was not obligated to pay lifetime health benefits since they weren't part of the regular union contract. (The dissenting judge, a Democratic appointee, penned a stinging dispute that accused Morrell of breaking its word).

Armed with that decision, which the U.S. Supreme Court upheld, Morrell made a change even more dramatic than the one it had originally proposed publicly. It wouldn't simply be reducing the retirees' health coverage. It would be eliminating the insurance altogether. Once again the union officials scrambled, this time meeting with local insurance agents to negotiate premium discounts, then holding meetings with Morrell's former employees in order to explain the available options. But the meetings were less about educating the membership than giving it a chance to vent—at the court, the company, and the union itself—about the loss of coverage. "We kept saying, 'they just can't do this, they can't,'" Audrey Sampson remembers. "But they did."

———

Clinton Baldwin had worked at Morrell for more than thirty years—retiring in 1985, the same year Lester Sampson did, and taking a job at a local bank. By then, Clinton, who had once served as the treasurer for the union, had observed firsthand the growing militancy of both the company and the union, concluding that it was time to get out. But no matter what happened at the plant, Clinton reasoned, his retirement benefits were secure. After all, he thought the company had made a promise to him, to Lester, and to all the other retirees. In fact, Clinton was so confident of those benefits that he declined health insurance from the bank when he started working there—in part because the bank, as a small business with just a few full-time workers, offered coverage that was considerably less comprehensive and far more expensive than Morrell's.

Clinton felt pretty good about the decision in 1991, when, at fifty-six years old, he had his first heart attack. The combined physicians' and hospital bill came to $30,000; Morrell's insurance covered everything but a $200 co-payment. Clinton didn't feel so good about his decision afterward, once he read in the newspaper that Morrell was scrutinizing its retirement benefits. Even though Clinton had always taken good care of himself—he'd had the first heart attack while taking one of his customary three-mile morning walks—he knew he was at risk of future medical problems. (Among other things, his family had a history of heart trouble.) So even before the union had exhausted its legal appeals to preserve Morrell's retirement benefits, Clinton had started looking for new coverage.

The trouble was that he couldn't find any. The bank's insurance carrier made coverage available only to new employees; since Clinton had turned it down on joining the company, he couldn't enroll afterward. (This rule is fairly common; it's another way insurers try to protect themselves from adverse selection, particularly among small-business clients.) Next Clinton tried to buy insurance from a private carrier directly. And then from another. He finally tried about ten or fifteen in all, he says. Not one approved him. As soon as the insurers heard he was a sixty-one-year-old man with a history of heart trouble, they turned him down.

Thus did Clinton Baldwin find himself uninsured when, in 1996, he had his second heart attack. He hadn't wanted to believe it was happening, putting in a full day's work at the bank before calling his doctor and, at the doctor's suggestion, going to the emergency room. This time the treatment cost $50,000—and Clinton owed all of it. When he explained his situation to the physicians' group, they reduced the bill (about $10,000 of the total) and allowed him to pay it back slowly, while continuing to provide him with discounted care. But Clinton had no such luck at the hospital. He says that the staff there rejected his offer of gradual payment, insisting that he pay the charges, nearly in full, within two years. Clinton and his wife couldn't make that up out of their savings. And after the second heart attack, Clinton wasn't supposed to work full-time. So they ended up selling their home—a split-level on a hill in an upscale Sioux Falls neighborhood. It fetched $70,000.

With some of the money left over, they purchased a mobile home to be placed on the outskirts of town, in a development behind an industrial tract of oil storage tanks and auto repair shops. The couple's concerns about health insurance subsided when Clinton, at sixty-five, became eligible for Medicare and (with the help of the money from the sale of the house) bought a generous supplemental policy. While the oil tanks weren't an attractive entrance to the neighborhood, the relentlessly upbeat Clinton preferred to talk about the view from his back porch—of serene farmland interspersed only with sheep and the occasional grain silo.

James Couch had a harder time with the change. James had worked at the operation in Sioux Falls for only a few years, moving there after Morrell closed the plant in Ottumwa, Iowa, where James had put in most of his thirty years. Like Lester Sampson, James had served in the navy. (It was the naval training center in Ottumwa, which operated from 1942 to 1947, that had first brought James—a native of Georgia—and many of his future coworkers to the midwest.) And, much like Lester, James had eventually worked in several jobs—finishing up in sanitation, where his job was to wash down the uniforms with chemical solvents. James was not among the company's better-off retirees; according to his son (James Jr.), James and his wife Marilyn "really had to watch their pennies," scraping together the money for a mobile home in Ottumwa. James also had a number of ailments by the time he retired, including emphysema and back problems, at least some of which—according to James Jr.—could plausibly have been blamed on the rigors of his work at Morrell.

Still, by several accounts James was happy enough in his retirement, taking particular delight in his daily fishing expedition. When the weather allowed it, he and some buddies from Morrell would spend a few hours fishing in the Des Moines River and a few hours more talking about it at the local Hardee's. (When the weather got too cold, they'd skip the fishing and go right to the coffee.) Then Morrell started talking about reducing retirement benefits—and James became, in turn, despondent and incensed. He worried about all the medications that he had been taking and that Morrell had been paying for; without the company coverage, he figured, they would cost him several hundred dollars a month. He also worried about how Marilyn, who had been out of work but was years away from eligibility for Medicare, would get insurance at all. And he complained bitterly that the company had turned its back on its workers. Friends told him there were other sources of coverage; among other things, James, because of his time in the navy, was eligible for services from the Veterans Administration (VA). But James became inconsolable. One day he made the ninety-minute trip to the nearest VA. The next day he took out a rifle and shot himself.

Over time, almost everybody heard about what happened to James Couch—and many came to blame Morrell for it. But what happened to him, and Clinton Baldwin too, was exceptional—at least in terms of severity. The Sampsons were far more typical. In 1995, Audrey, a few years younger than Lester, was in relatively good health and was able to buy health insurance from Blue Cross, which promised to cover the majority of her health care needs until she turned sixty-five. Lester's previous heart attack meant that he probably could not have gotten such comprehensive benefits on his own. At the very least, any coverage he might have obtained would almost certainly have excluded coverage for anything related to his heart. But since Lester was sixty-five, he didn't need to find full coverage. Medicare would take care of the basics. He just needed supplemental coverage to fill in the gaps, and he was able to purchase that.

Then again, the Sampsons had given up $100 a month worth of pensions in order to get the promise of retirement health benefits— money that wasn't restored even after Morrell stopped providing the insurance. And because the Sampsons believed that Morrell would take care of them, they'd each declined a separate offer to get retirement health insurance—Audrey through her job at UPS, Lester through a job he'd taken in the school system after leaving the meatpacking business. (As with Clinton Baldwin and the bank, each offer was available only at the start of employment; having turned down the benefits once, the Sampsons were no longer eligible to get them.)

Now the Sampsons were looking at purchasing extra coverage, for about $400 a month. And those policies still had large gaps, particularly when it came to outpatient prescription drugs—a problem that became severe once Audrey, who'd also been a smoker, started developing breathing problems. The union had gotten some Sioux Falls pharmacies to agree to discounts for Morrell's retirees, but these discounts reduced the price of the drugs by only a small fraction. The Sampsons figured that buying their drugs locally would cost them hundreds of dollars more a month—which, combined with the supplemental premiums, was not something they had factored into their budgeting for retirement.

For a while, the annual trips to the south provided a convenient solution. About an hour's drive from Phoenix, where the Sampsons camped, was the city of Nogales, Mexico. And just a few steps across the border were pharmacies that sold name-brand drugs, or generic versions, at deeply discounted prices. (Some of the stores had physicians on-site, to write the prescriptions; others were happy to fill prescriptions on request.) The stores were there to meet a growing demand by Americans, most of them seniors frustrated with the high cost of drugs in the United States. Indeed, by the late 1990s entrepreneurs were running regular bus services between countries—and doing a brisk business—under names like the "Prescription Express." One survey eventually found that 5 percent of all seniors in America had bought drugs from Mexico or from Canada, which had a similar cross-border trade. The Sampsons were among them, and for the first few years after Morrell's coverage vanished they thought the Mexican drugs were a godsend.

But their truck wore down from all the traveling. And so, eventually, did the Sampsons. Instead of going south each winter, they would have to find an apartment in Sioux Falls. And that meant finding money to pay for it. They started by selling the truck and trailer, which, after all, they didn't need so much anymore. But with so much mileage, neither fetched much money. That left just some meager bank savings, pensions, and Social Security—hardly enough to cover the rent for the apartment, particularly if the Sampsons wanted to hold some money in reserve, in case one of them ended up needing long-term nursing care (a distinct possibility, given their ailments). Reluctantly, they turned to their last option: they sold the cabin on Lake Madison.

The sale itself was easy, and the money it produced padded their savings, although they began drawing the savings down again almost immediately. The hard part of the transaction was the emotional impact. The apartment they found was in a complex that sat behind a car wash on one of the busiest commercial thoroughfares in Sioux Falls; where once the Sampsons could look out on wilderness and the lake, they now saw concrete and strip malls. And just as Audrey had lost her garden, so Lester had lost his fishing porch. When the couple first moved, Audrey noticed that Lester seemed depressed and urged him to try fishing on the Big Sioux River, which ran through Sioux

Falls. But this was a poor substitute and, as it happened, an impractical one. To fish on the river, you had to stand on the banks or get into a boat. Neither option was really feasible for a man who, like Lester, had recently become tethered to an oxygen line that followed him wherever he went.

————

Morrell's decision to cancel health insurance for retirees drew national attention. And critics were quick to point out that, even as companies like Morrell were supposedly struggling, their parent companies were making good money. "It was an unbelievable demonstration to me of the callousness that sometimes occurs to employees who have dedicated their lives to a company," said Thomas Daschle, the Democratic senator from South Dakota. "How could a company cancel lifetime benefits? How could the courts let them? And the Morrell retirees were not alone."

On that last point, Daschle was correct. As a series of private surveys and government reports confirmed, the number of employers offering retirement insurance was declining substantially. And in a few high-profile cases—many of them involving the struggling steel industry—large companies had canceled (or would go on to cancel) benefits that current retirees were already receiving, causing some retirees to deplete their savings and others to go without recommended medical care. But there wasn't a whole lot the politicians could do to stop Morrell and other industrial behemoths from dropping retirement insurance. Even if the companies were going back on a promise—and many insisted that, legally speaking, they weren't—they were simply doing what the global economy demanded of them: they were trimming their labor costs in order to compete with companies producing cheaper goods.

So as anxiety over the rising cost of drugs grew among retirees, politicians began talking about something they could do. They could expand Medicare to cover prescription drugs, doing for all seniors what Morrell had once done for Lester and the rest of its retirees. Polls found that the public generally supported this idea. Even younger people favored it, perhaps because they could envision themselves needing help someday—or because they knew that without help from the government, the financial burdens on their parents and grandparents would eventually fall to them, much as had happened in the days before Medicare.

It was the same dynamic that had existed in the early 1960s, in more ways than one. Once again, reformers chastened by the failure of proposals for universal health care—in this case, Clinton's plan rather than Truman's—were focusing on the needs of the elderly as a way to pass health care legislation.

But the debate of the 1990s over how to help the elderly would play out differently from the debate of the 1960s. Precisely because the elderly already had Medicare, the case for helping with drugs raised eyebrows among some pundits and politicians who believed that senior citizens, on the whole, were already receiving too much government assistance. If the government was going to get into the business of filling in Medicare's gaps, the argument went, it had to do so in a way that delivered benefits only to those in clear economic need—and only as part of a broader package that reconfigured Medicare, if not the entire welfare state, to restrain its growth. In the eyes of these critics, the notion of simply adding a drug benefit to Medicare, for which every senior citizen would be eligible, was—as the *Newsweek* columnist Robert Samuelson put it—"mostly a shameless competition for retirees' votes."

Implicit (and, in the case of Samuelson, explicit) in this view was the idea that Medicare would work best as a welfare program—a government safety net that ought to be as minimal as possible, filling in the holes of the private market for health care and doing nothing more. It was the very same argument made by critics of Medicare in the early 1960s. When the AMA and other industry groups were fighting Kennedy, Johnson, and their allies, they too suggested creating an alternative program that would cover only the very poor and, perhaps, the medically uninsurable. At the time, that argument had failed, largely because it had become so readily apparent that the private insurance market would never provide even the majority of seniors with adequate health insurance—and that providing the elderly with access to medical care was a job for which the government was best, and uniquely, suited.

But by the 1990s, renewed faith in the private sector had displaced faith in government, not only in Washington but in the public at large. In the thirty years since the creation of Medicare, the proportion of Americans who "trust[ed] the federal government to do what is right most of the time" had fallen from 69 to 23 percent. And that sentiment—the same sentiment that had helped kill Clinton's universal health care plan—had a profound effect on the two Medicare reforms that would follow.

The first came in 1997 as part of the budget deal, when Clinton and the Republican Congress agreed to create "Medicare + Choice." The initiative vastly expanded the number of private managed care organizations offering coverage to seniors (previously, only a few had offered it) while encouraging seniors to sign up. The hope was that HMOs and other forms of managed care would reduce the costs of Medicare, just as they had done for employers in the private sector, while offering more benefits and promoting better medical care. Initially, they did just that—and became a vital source of prescription coverage for many seniors who needed it.

But the features of managed care that made it unpopular with everybody else, such as its limited networks of providers and its bureaucracy for approving individual procedures, were particularly threatening to seniors, since they tended to use the most medical care and to have the longest-standing relationships with particular physicians. And while studies on the quality of care these HMOs provided to seniors were generally inconclusive, the verdict on their finances was quite clear: far from saving the taxpayers money, managed care in Medicare was actually wasting it. Per person, government paid HMOs almost as much to cover senior citizens as it spent to cover them directly through traditional Medicare. But a series of studies subsequently revealed that the HMOs were largely attracting healthier seniors, either because only those seniors were willing to take their chances with managed care or because the companies were actually targeting those groups deliberately. In either case, the government was significantly overpaying the HMOs.

When the Clinton administration (again, with bipartisan support) promptly reduced those payments to bring them more in line with actual costs, the HMOs started pulling back. Some cut back on prescription drug coverage, which for many seniors was the whole reason for joining managed care in the first place. In 2003, less than half of Medicare HMOs were offering more than $750 in total prescription drug coverage, a number that represented one-fourth of the average beneficiary's total drug costs. (Back in 1999, 79 percent of the HMOs had offered that much coverage.) Others simply dropped out. The problem was most acute in rural and less populated areas, where seniors frequently had no managed care options at all; this frequently meant that those without employer-provided retirement benefits had no prescription drug coverage.

The health insurance industry had now failed senior citizens twice—once in the 1940s and 1950s, then again in the 1990s. But the blind faith in private insurance for the elderly didn't end with Clinton's presidency. On the contrary, it gained an even more committed advocate in President Bush, who made it the centerpiece of his health care agenda during his first term. In 2003, Bush signed into law the

Medicare Modernization Act—which gave seniors their long-awaited prescription drug benefit, but delivered it through private insurance carriers rather than the government. The not-so-secret agenda of this measure was to upend the entire Medicare program, transforming it into a system of competing private insurance companies—a goal conservatives had been pursuing openly since they took over Congress in 1995. But in his public rhetoric, at least, Bush didn't stress this point, instead focusing on how his plan would help seniors by giving them greater "choice" and promoting efficiency. "The days of low-income seniors having to make painful sacrifices to pay for their prescription drugs," Bush declared, "are now coming to an end."

———

Lester and Audrey Sampson, like millions of seniors, weren't so sure. On a blustery December day in 2005 that was frigid even by South Dakota's standards, they sat inside their electrically warmed apartment, frustrated and overwhelmed by the new program. Like most seniors, they didn't really crave choice. They wanted protection—the kind that old-fashioned, government-run Medicare had always given them for doctor and hospital expenses. But they were having a hard time finding it in Bush's plan.

At this point, the problem wasn't so much Lester's drugs. He, too, had discovered the VA, which filled all his prescriptions for just $125 a month. Audrey, though, was another matter. She was taking ten pills, among them a drug called Tracleer that helped her cope with pulmonary hypertension, a relatively rare lung disorder that made her short of breath and, if untreated, could kill her. Having just won FDA approval in 2002, Tracleer was an alternative to more established treatments that had more severe side effects or provided less relief from the symptoms. But the studies showing its effectiveness had been quite convincing, so much so that private insurers routinely covered it for their working-age beneficiaries who met the basic clinical criteria. (The main problems were potential liver damage and infertility; neither of these was a concern for women of Audrey's age.) And that was no small matter, since in the United States the retail value of a bottle of sixty Tracleer pills was $3,000.

Audrey didn't need a study to tell her the drug worked. She knew from the moment she began taking it. What she needed to know was which of the new Medicare prescription drug programs would cover it, since to pay for it she had been relying on a state program for low-income seniors—a program that was expiring because the new Medicare program was supposed to take its place. But figuring this out was no easy task. The government had distributed booklets to everybody eligible for the new Medicare prescription drug benefit, offering details about the different coverage options so that beneficiaries could make a decision by January 1. But the mountain of detail was enormous and confusing. Each plan covered different drugs, with varying co-payments. In some cases, you paid more if you used fewer pills; in others, you paid the same rate no matter how many you used. Not every drug was listed; to find out about relatively rare drugs, you had to contact the company directly. Most bewildering of all, an insurer could decide to drop coverage at any time, but beneficiaries were allowed to switch plans only once a year.

The Sampsons weren't the only ones who were confused. For at least a year, experts had warned that seniors would have trouble figuring out the new system and that some of them—particularly the poorest and sickest ones—would struggle to get their drugs once the program actually started. And in the final weeks of 2005, articles about seniors baffled by the program popped up in newspapers all over the country, just as the experts had predicted. The administration continued to brush off warnings, predicting a smooth transition. But then the program actually began—and chaos ensued. Tens of thousands of senior citizens, maybe more, showed up at pharmacies only to discover that they didn't have working drug coverage; many would leave without their drugs. In Maine, a hotline for confused seniors logged 18,000 calls in one day. "We had dialysis patients who were not getting medicines, pharmacies on hold for 60-plus minutes, some plans closed for the holiday," one state official reported. "One man called me—he and his wife were on 15 medications. . . . He went in for 15, and he left with one."

The contrast to the implementation of the original Medicare program some forty years before was striking. To address fears that hordes of newly insured seniors would overwhelm hospitals, officials of the Johnson administration had decided that coverage would begin in July, when hospitals were traditionally least crowded. And although their own calculations showed that worries about overcrowding were unfounded, they drew up emergency plans for moving patients to overflow facilities—going so far as to line up military helicopters in Texas for speedy transfers. The architects of the law, both at the White House and in Congress, had also decided against offering seniors a series of options for coverage, recognizing—as the Bush administration had not—that in some cases choices could be more debilitating than empowering. Most important, they were dealing with seniors directly through the government, not through a series of private insurance intermediaries, using the knowledge and resources honed through the Social Security system. The result? Under the headline "Medicare Takes Over Easily," a *Washington Post* writer described the program's first day as "a smooth transition, undramatic as a bed change." Three weeks later, a *New York Times* headline affirmed that "Medicare's Start Has Been Smooth."

It would take a few months before anybody could plausibly use the word "smooth" while discussing Bush's Medicare drug plan. And, even after several months, only about half of the nation's senior citizens had enrolled. (By comparison, on the very first day that Medicare went into effect in 1966, nearly 90 percent of eligible seniors had already enrolled.) Still, once the Bush administration worked out the kinks in the computer systems, the drug program seemed to be working for those who had figured out how to take advantage of it—a category that included the Sampsons. Audrey never did find a insurance carrier that provided adequate coverage for Tracleer; most carriers either didn't cover it at all or covered it only partially, requiring thousands of dollars in out-of-pocket spending that she and Lester didn't have. With some determined searching, however, Audrey finally found a private organization that provided assistance to people like her. That, plus the Medicare stand-alone policy she chose (a Blue Cross plan), was giving her the coverage she needed.

Future generations, though, had reason to remain concerned, because the long-term questions about the new drug benefit were less about its workability than its sustainability. One reason the Bush administration had insisted on channeling the benefit through private insurance was that the insurers would not be able to dictate price reductions to the pharmaceutical industry the way the government could if it were managing the benefit directly. But that made the program far more costly—to the tune of hundreds of billions of dollars, by some estimates. And since private insurers had demonstrated during Clinton's presidency that they would not stay in the Medicare business if they couldn't make a hefty profit, the Bush administration proceeded to lard up the new law with additional subsidies to the insurance carriers.

In a sense, these decisions represented nothing more than good old-fashioned influence peddling; the pharmaceutical and insurance industries were generous funders of Republican campaigns. But the shape of the Medicare drug benefit also reflected a strong ideological aversion to big government: the Bush administration was so convinced of the private sector's inherent efficiency that it was willing to waste billions of taxpayer dollars to prove it. Of course, the more expensive the drug benefit became, the more insistent calls for reining it in were bound to become, just as such calls had been in the 1970s. Medicare had finally expanded to meet society's growing needs, as its original architects always hoped it would. But, at the behest of those who never really believed in the program, it had expanded in a way that undermined its effectiveness—and jeopardized its long-term survival.

FIVE

LAWRENCE COUNTY

Wanda Maldonado watched as Ernie tossed and turned that October night in 2005, trying desperately to get comfortable. She knew that her husband had overcome many struggles in his life, from the time he first arrived from Puerto Rico as a fourteen-year-old, with just $5 in his pocket and speaking virtually no English, to the time he moved with her to her rural Tennessee, at age fifty-five, establishing a successful automobile painting business out of nothing. But Wanda also knew that Ernie, now sixty-eight, had never faced an adversary quite like this one. He was fighting a war of attrition with disease. And disease was winning, one organ at a time.

Ernie needed artificial oxygen because his lungs, stiffened by years of smoking and inhaling industrial fumes, had lost their ability to expel air from his body. He could not walk without pain because neuropathy, which he'd gotten from lead in the paint, had damaged the nerves in his legs. His arteries were clogged, straining his heart; plus he had developed diabetes, weakening his immune system. With his appetite gone, he was wasting away. In the last few months, he'd dropped ten, twenty, and then thirty pounds. His eyes sank into his skull. The olive tint drained from his face.

In health, Ernie had been famous for stubborn independence and a mischievous sense of humor. And perhaps because he hated being an object of sympathy, he could still summon the strength to put on a show when friends came to visit, trying to mask his pain and distract attention from his diminished appearance by teasing his company or telling dirty jokes. But when he and Wanda were alone, as they were this night, he'd drop the pretense. She understood how much he was suffering. And this day, she had realized early on, was going to be one of the bad ones.

It had started in the morning, when Ernie had a bout of diarrhea that left him feeling sick to his stomach all day long. He had also complained that he was having unusual difficulty breathing. "I had tried to get him to let me take him to the hospital, to the emergency room, but he wouldn't," Wanda later explained. "He said, 'If I ain't better in the morning, I'll go to the doctor.'" She agreed, checking on Ernie regularly all the way until four a.m., when he finally rolled over and fell asleep. Relieved that he was resting, she retreated to their bedroom next door, making sure to turn on the baby monitor she kept in the headboard, just in case Ernie woke up and needed something right away.

Despite her exhaustion, Wanda could sleep just a few hours. At seven a.m., she found herself awake and, although semiconscious, going through her morning routine: pouring a cup of coffee, letting out the cat, and smoking a cigarette on the porch. Something felt wrong, though, and halfway through the coffee she decided to check on Ernie. She found him in bed, still turned to one side, in exactly the same position as he was when she had left him. "Ernie?" she called. "Ernie?" she said again, this time more insistently. "Ernesto?" she said finally, but still no answer. Nervously, she put her hand on him and realized he wasn't breathing. She picked up the phone and called 911. But she was just going through the motions, she realized later. "I knowed he was gone."

Wanda wept, though more in grief than shock. She had understood that this day would come, probably sooner than later, just as she understood that she should be grateful for the last few years, which the couple might not have had together were it not for the intensive, scientifically advanced medical treatments Ernie had available to him. Doctors had saved Ernie from death once already, years before, when he suffered a potentially fatal heart attack. Since then, regular therapy for his various ailments plus state-of-the-art prescription drugs had not only kept him going but also alleviated his suffering. Steroids had softened the passages of his lungs, so that they could expel more air; anticoagulants had prevented his blood cells from clotting, so that his heart could continue to circulate blood through his plaque-hardened arteries. To keep the diabetes at bay, he took pills to maintain a steady blood sugar level and, as necessary, injected insulin. While his ability to be active had steadily declined over time, he had still managed to get out regularly for most of the last few years, whether it was to go fishing in a nearby stream or driving in the country with Wanda and his grandson, a boy of elementary-school age, whom Ernie delighted in letting play navigator.

But if medical science had made this time available to Ernie, it was a government-provided insurance plan that had put it within financial reach. That program was Medicaid. Since its inception in 1965, Medicaid had steadily expanded, covering an ever larger share of the population to accommodate the growing number of Americans who needed assistance in affording their medical care. Tennessee's version of the program, called TennCare, was among the most ambitious: by the late 1990s, it was covering nearly one-fourth of the state's population. What's more, the insurance itself was as comprehensive as anything in the private sector, covering a wide range of doctors' services and hospital services, plus prescription drugs, while requiring only token out-of-pocket spending.

Such generosity would not last, however. In 2004, Tennessee's governor, citing budget pressures, rising health care costs, and inefficiencies within TennCare itself, scaled back the program. The early cuts were modest; the subsequent ones were severe. Some people would end up losing their TennCare altogether. Others, like the Maldonados, would stay in the program but lose crucial benefits. By the fall, Tennessee was covering just a few of the Maldonados' prescriptions, leaving the couple to beg for medicinal handouts at the local clinic where they got their care—or, when samples weren't available, to simply go without some of the drugs. It was not long after Ernie began cutting back on his medications that he began his steep decline. And it was just a week after Ernie had stopped taking Plavix, the blood-thinner that prevented blood clots, that he died from what doctors later said was a heart attack.

Wanda understood that there was no way to determine whether the rationing of medications had hastened Ernie's death. In fact, she would later say with resignation, "God has our numbers written before we are born." But Wanda also couldn't escape the idea that, were it not for the reductions in TennCare benefits, Ernie might have lived for a little while longer—or, at least, lived in less discomfort. And over time, she began to ponder something else. Relative to other TennCare beneficiaries, Wanda figured she and Ernie had been well-off. What, she asked, was happening to all those other people who had depended on the government for their health insurance?

It was a very good question. And not just in rural Tennessee.

Despite all the anxiety about paying doctors' bills and hospital bills during the 1920s, helping the poor was barely an afterthought. In fact, the solution to unaffordable medical care that grew out of that era, Blue Cross-style insurance, did very little to help Americans with the lowest incomes, since you couldn't even get the coverage unless you had a job or money to spare, or preferably both. And although a few programs run by the government during the Depression provided extra assistance for medical care, they were financially inadequate and didn't last past the early years of the New Deal. So over the next two decades, while private insurers were making the miracles of modern medicine easily accessible to most of the middle class, people with very little money were forced to rely on town doctors, big-city public hospitals, and private charity—a situation that, as one expert explained in the 1950s, with considerable understatement, "leaves much to be desired."

Ironically, it was primarily those groups opposed to government-run health care that finally produced serious talk of having Washington do something to help low-income Americans get health care. With momentum growing for a program to help all of the nation's elderly, organizations representing physicians and the insurance industry began clamoring for what seemed to them a less offensive alternative: a program that would help only the elderly with severe medical and economic needs. Insurers, after all, didn't want to cover these people anyway: they represented high medical risks yet lacked the money to pay for appropriately priced coverage. And, given the relatively narrow scope of such a program, it would be unlikely to interfere too much with the practice or economics of medicine. With support from the Eisenhower administration, which shared the medical industry's aversion to government interference in health care, Congress in 1960 passed what became known as the Kerr-Mills Act.

Kerr-Mills conformed to long-held ideas about which Americans deserved help and which ones didn't, limiting its assistance exclusively to people who were both elderly and "medically indigent"—meaning that the costs of paying for their medical care could overwhelm their incomes. And even though the federal government was putting up the money for Kerr-Mills, the administration and design of the program were left largely to the states. That was consistent with another long-standing American belief—that welfare is primarily a local issue—first established by the Elizabethan poor law of 1601 and carried to the New World by the original colonial settlers.

Backers of Kerr-Mills had predicted that it could guarantee every senior citizen would have "an adequate medical program." It did nothing of the sort. Even after the initiative went into effect, both the elderly and the poor continued to struggle, strengthening calls for a more massive intervention—something like what Johnson and the Democratic Congress were pushing with their Medicare proposal. The AMA and its allies made a final stand of sorts, suggesting that the government strengthen rather than replace Kerr-Mills. But that's when the supporters of government-run health care outmaneuvered them: rather than simply steamroll their opponents, they co-opted them—or, at least, their arguments. The new health insurance act would have a universal program for the elderly *and* a targeted program for the needy, old and young alike. The latter would become the Medicaid program.

When Johnson signed the Medicare-Medicaid bill into law during that ceremony at the Trumans' home, he mentioned Medicaid only briefly—which was indicative of how (relatively) little thought he and the rest of Washington had really given the program. But the architects of Medicaid had quietly packed it with enormous potential. It filled in the gaps of Medicare not just for the elderly with very little money, but also for people who were blind or had permanent disabilities. (Today 40 percent of the program's money is spent on these groups.) It extended health coverage to everybody receiving cash assistance through welfare, primarily single mothers and their children. And despite its very extensive list of services covered, it involved little or no cost sharing. Like Kerr-Mills, it would be administered primarily by the states in accordance with federal guidelines. But although the guidelines allowed states considerable leeway to shape the program, the states would now have much more incentive to be generous, because the federal government was putting up more of the money—much more than it had put up for Kerr-Mills.

It was a remarkably ambitious reach, made possible by one key political ingredient: stealth. Unlike Medicare, which had been crafted after literally years of testimony, argument, and (eventually) negotiation among the many interested parties, Medicaid was slapped together in the final weeks of legislative haggling. And even then, Medicare itself was still attracting most of the attention, allowing Medicaid to escape the intense scrutiny of officials and interest groups that might have seen fit to tinker with it.

But precisely because nobody bothered to fight against Medicaid, nobody had bothered to fight for it, either. Its supporters never went through the effort of making its case publicly—of convincing the voters that spending so much of their tax money on needy people was a worthwhile endeavor. That would render the program vulnerable once it got the critics' attention—which, all too predictably, happened almost right away.

According to the original projections, adding Medicaid to the Medicare law would cost just an additional $238 million. But the expenses blew past that figure after just six states had launched their programs. With political momentum already turning against Johnson and the programs of his Great Society, Congress acted quickly to rein in the new initiative, limiting eligibility to people below or near the poverty line. New York, a state with a generous social welfare tradition and a progressive Republican governor—Nelson Rockefeller—had proposed opening its version of Medicaid to so many people that nearly half the population could have qualified. Had that effort succeeded, New York would have had close to universal health coverage. Now the state would have to settle for a far less generous program—and continue to have large numbers of uninsured residents.

The next two decades would follow much the same story line. The demand for Medicaid kept growing, as more people either lost their private health insurance or found that their insurance didn't cover all their needs. In response, officials like Congressman Henry Waxman kept pushing to expand the program, albeit quietly. Even as opponents of government-administered health insurance were carping about the program's costs, Waxman and his allies were slipping expansions into the budget reconciliation process, so that these expansions could sail through Congress without the usual scrutiny or wrangling with special interests. (The expansions typically cost very little the first year, to avoid causing controversy, then increased in cost afterward—a trick that critics called the "Waxman wedge.") Every year between 1984 and 1990, Congress loosened Medicaid's eligibility requirements for pregnant women and children, bringing benefits to 5.5 million people who would not have had them otherwise.

These expansions reflected not just a sense of moral obligation toward the poor and a recognition that their needs were growing, but also an understanding that people with low incomes had special medical needs. They had a less healthy diet, had less healthy living habits, tended to work in environments that were more physically hazardous, and had the least consistent contact with regular preventive care. They needed not just better insurance but general public health intervention, in order to reconnect them with modern, mainstream medicine. And particularly when it came to prenatal care, the early investment actually seemed to pay off economically, since it reduced the likelihood of low-birth-weight babies who were at greater risk of serious, costly birth defects.

But the political downside to this strategy was the same as it had been when a similarly unheralded effort had created Medicaid in the first place. The program's ambition was exceeding its popular mandate. That made it an easy target for those who wanted to reduce the reach of government, starting with Ronald Reagan in 1981. From the beginning, Medicaid, like Medicare and Social Security, had been an "entitlement": if more people were meeting the program's eligibility guidelines, then the law required that states give them coverage and that the federal government pay its share of the additional costs. Reagan wanted to make it a "block grant," under which the federal government would simply decide on a fixed contribution to the states every year. Since Medicaid always seemed to grow faster than the projections, this probably meant that the states would find themselves without enough money to support the program. The states would then be forced to cut it, unless they could find new sources of revenue on their own. When even some Republicans complained, fearing that their own states would suffer, Reagan's plan failed. But his administration still managed to enact a sweeping cut in Medicaid as part of his budget for 1982.

The end result of this back-and-forth was an affirmation of the political message first sent in 1969, when Congress reined in New York's effort to make Medicaid a vehicle for universal health care: the program would serve only those Americans in the most extreme need, under the strictest criteria of eligibility. And this was just the way the champions of private insurance liked it. If government had to get involved in providing health insurance at all, it should serve only a small, particularly impoverished group of people that private insurers didn't want to touch anyway. This arrangement would reduce the political pressure for broader government intervention without getting in the way of the rest of the insurance market.

Still, there was a problem. A program that could cover only such a narrow group of beneficiaries was bound to leave out some people who legitimately needed help.

As a young man in Puerto Rico, Ernie Maldonado had a notorious propensity for getting himself into trouble. In the early 1950s, according to Wanda, he had gotten an older girl pregnant and embarrassed his family—particularly his father, who was then running for mayor of the town where they lived. When his father lost that election, Ernie ran away, worked to earn some money, then spent the money on a one-way ticket to Miami. On the day when he arrived, he made his way to a coffee shop and tried, with some difficulty, to buy a ham sandwich. He finally got it, but only after paying the $5 that was then his entire net worth. He stayed with family friends and eventually, through them, became a migrant farm worker. But he got fired from his first job after picking tomatoes the wrong way—and then repeating some of the English curse words he'd overhead in conversation.

But Ernie also had an industrious side. He eventually learned English by working with a friend who wanted to learn Spanish; the two used comic books as their guides. He also read widely, showing a particular interest in medical books. (For a while he dreamed of becoming a psychiatrist. He gave the idea up when nobody in his family took it seriously.) Eventually Ernie found work painting automobiles, a trade he'd first learned in Puerto Rico, where his family had a car servicing business. He quickly gained a reputation for unusual precision, and in the mid-1980s he was hired by a company in Jacksonville that built and outfitted heavy equipment for ports, ships, and other industrial purposes.

By this time Ernie had moved in with Wanda. The two had actually met many years before, when Wanda had first come to Florida. At the time, she hadn't particularly cared for Ernie. But since then she had returned to Tennessee, where she won custody of the children from her first marriage, and had brought the children to Florida. And during that second stay she suddenly found herself attracted to Ernie, whose own marriage was just ending. "He had been hurt and I had been hurt," Wanda would later remember. "First it was just an easy-going relationship. Then we realized we both had what we wanted and let down the wall."

At first, the two avoided talk of marriage. But then one day Ernie's boss nearly died from a heart attack—surviving, according to Ernie's retelling, only because his wife had been on hand to give the doctors important information. On hearing that story, Ernie came home and told Wanda it was time for them to get married. She agreed and wanted to know if he had a date in mind. It turned out he did. On a Wednesday afternoon in 1991—just two days after Ernie popped the question—the two wed. The haste hardly seemed to bother Wanda, who felt as close to Ernie as she had to anybody in her entire life. "He made me," Wanda remembered. "And I knowed him, inside and out."

Ernie's work was going well, too: he earned special certification to work on navy equipment, since he could meet the precise specifications for paint thickness, making him a particularly valuable—and valued—employee. For the first time in their lives, Ernie and Wanda had finally found some equilibrium.

But then Ernie started to have trouble with his legs. Working on port equipment required him to do a lot of climbing; some of the cranes he painted for the navy were so large that whole trucks could fit under their bases. The climbing had put a huge strain on his knees; and he was also starting to suffer the effects of neuropathy—something he'd gotten from the lead in the paint—although he didn't know it at the time. Ernie's doctors told him he couldn't continue to place such physical demands on his legs. And while his supervisor offered to arrange a desk job for him, Ernie politely declined, thinking he'd go stir-crazy. He figured he could still paint. He just had to do so under less strenuous conditions. So he suggested to Wanda that they open their own auto body business. And instead of staying in Florida, he suggested, they could move back to Tennessee—to Wanda's home-town, Lawrenceburg.

Lawrenceburg, a town of a just over 10,000 people, sat in the rural country southwest of Nashville, just thirty miles north of the Alabama border. It had plenty to recommend it as a place to live: close proximity to the mountains and outdoors; affordable housing; and, in Wanda's case, family. It also had a certain historical novelty: it is said to be the birthplace of gospel music, and it was once the home of Davy Crockett, who represented the town in Congress.

But Wanda's enthusiasm for coming home was dampened by her concerns about Ernie's business plan. He was an intelligent, articulate man—even though his schooling had ended after the third grade—who had assimilated into American culture with remarkable ease. But he still spoke with a thick Puerto Rican accent, which wasn't heard much in that part of Tennessee. And the couple had virtually no savings to their name. Starting up a new business and acquiring customers was bound to be difficult.

Wanda's solution was to put herself in charge of Ernie's business promotions. While Ernie was back in south Florida for a few weeks, helping friends and family whose homes had been wiped out by hurricane Andrew, Wanda went on a tour of the area's car dealers. She drove in a pickup that Ernie had painted, showcasing it as proof of his skills. The pitch worked. The original plan had been to work five days a week, from eight until five. Within three months, according to Wanda, they were working all seven days, sometimes as late as nine or ten o'clock at night. Business was so good that they ended up hiring an extra employee.

Soon the money coming in from the business allowed Ernie and Wanda to move from a two-bedroom apartment situated on a major highway to a four-bedroom home (still a rental) tucked behind a church. It was a much-needed shift because Ernie's children from his own previous marriage were now living with them part of the year, as were some of Wanda's. The house still had only one bathroom—the waits seemed interminable, sometimes—but the extra bedrooms at least meant that everybody had plenty of sleeping space.

What their money could not buy the Maldonados was insulation from high medical expenses. Back when Ernie was working in Jacksonville, they'd gotten insurance through his job. But that was group coverage. If Ernie and Wanda Maldonado wanted coverage now, they'd have to find either individual policies or something for small businesses—both of which would require going through medical underwriting. Because of Ernie's knees and Wanda's heart murmur—a problem that had been diagnosed about ten years before—the only insurance available to them would have been extremely expensive and full of exceptions for many of the conditions likely to incur medical bills.

Neither of them was particularly happy about this, particularly since Ernie was developing new health conditions, most serious among them a deep, persistent cough. Wanda didn't know exactly what lay in store for the two of them, but she could see that he and eventually she were going to need more medical attention. And that meant they would need some kind of insurance.

In 1992, Ned McWherter was in his fifth year as governor of Tennessee. And, like most of his counterparts across the country, he had a policy crisis on his hands. Higher unemployment from the recession that had begun in the late 1980s combined with rising health insurance premiums meant that the number of Tennesseans without health coverage was soaring. But Tennessee's Medicaid program—the state's main instrument for helping these people—was already de-

pleting the treasury, to the point where spending on health insurance
for the needy was about to overtake spending on public education.
Tennessee could try to cover the added expense by raising taxes, but
that would be both bad economics (raising taxes during a recession
might retard economic growth even further) and bad politics (Tennessee is a relatively conservative state that doesn't even have an income
tax). McWherter needed an alternative. And, as in almost every other
state, he quickly found one in managed care.

By the early 1990s, HMOs had already proved adept at holding
down the cost of medicine in the private sector. Their potential to do
the same in the public sector was, naturally, what made them such an
obvious remedy for the states' Medicaid crises. But if economics had
made the switch to managed care seem necessary, many experts hoped
it might yield medical benefits, as well. Even though Medicaid offered
a wide range of benefits, including the kind of preventive screening
measures that low-income people so desperately needed, studies had
shown that large numbers of beneficiaries never took advantage of
these services.

One reason so many Medicaid recipients never took advantage of
the services available to them was that many were simply disconnected
from the health care system: they used medical care sporadically, in
emergencies, but were far less likely to schedule regular checkups or
even to have a regular doctor. Managed care, as envisioned by its
enthusiasts, was the perfect antidote, since it would link each beneficiary with a primary care provider and network, establishing a lasting
relationship that would grow over time. Not everybody on Medicaid
would respond, to be sure; but for those who showed initiative, Medicaid would offer an opportunity to get regular medical care, the kind
many well-insured people took for granted.

McWherter was sold—though he was careful about how he unveiled his proposal. Aware that the public might be wary of such a
massive shift, he figured that people would go along with it if he could

use the savings to cover more beneficiaries. And that's just what he proposed. TennCare would not only cover the groups traditionally covered by Medicaid: children, single mothers, and low-income elderly people. It would also offer state-subsidized coverage, with modest premiums, to people who could not get insurance even though they were too wealthy, or of the wrong demographic profile, to fit those categories. People with incomes as high as 400 percent of the poverty line would be allowed to buy in. When fully implemented, the plan would cover one-fourth of the state's population.

Ernie and Wanda were among those who, despite not qualifying for Medicaid in the past, would now be allowed to buy coverage. And they jumped at the opportunity. For a premium of only $30 per month—a rate impossible to find on the open market—they got full doctor, hospital, and prescription coverage, with co-payments of just $1 or $2 for some services. The coverage came not a moment too soon. Shortly after going on TennCare, Ernie suffered a heart attack. He had to be transported by ambulance to the nearest large hospital, in Columbia, where doctors performed open-heart surgery. The doctors also determined that his cough was an early stage of emphysema. They treated it with steroids, which made breathing possible, but taking steroids over an extended period of time led to other medical problems, affecting everything from his digestive tract to his kidneys—and, most likely, contributing to the development of his diabetes. The treatment for all these conditions involved yet more visits to doctors and, inevitably, more drugs.

During the heart attack, Wanda later recalled, Ernie was typically—even absurdly—upbeat, cracking jokes while the emergency room staff worked frantically to save him. He returned to work quickly and insisted on continuing, even after he had begun to develop severe back pains that came from either the neuropathy or the strain from trying to breathe through the emphysema. His solution was to simply lie flat on his back for hours at a time, working on the undersides of cars, moving from one vehicle to the next on a wheeled platform so that he wouldn't have to try to get up.

Eventually, though, the pain became too severe even for Ernie: he sold the shop and filed for disability benefits. That left him and Wanda living on just $500 or so a month. But this was enough, since TennCare took care of all the medical bills. In fact, even with all the extensive medical treatment Ernie was receiving, he and Wanda found they still had money for the small luxuries in their lives, like an occasional dinner at the Ponderosa steak house, where Wanda's daughter worked, or Long John Silver's.

———

"TennCare was great," Wanda would later say. But elsewhere in Tennessee, people had far less sanguine reactions. The early months of the program had been chaotic, as tens of thousands of people found themselves randomly assigned to plans without enough doctors. And once those glitches subsided, more persistent problems emerged. The HMOs were supposed to realize their profits (or, in the case of the nonprofit HMOs doing TennCare business, to balance their books) by providing more efficient medical care. But particularly early on, some had applied a simpler strategy—one recognizable to anybody familiar with the history of private insurance and the way it had traditionally competed: they avoided enrolling people likely to run up large medical bills. In a story for the *New York Times*, an adviser to one TennCare HMO explained how he advised its recruiters: "Be observant. If you see a physical problem, you don't need to sign those people. If you see someone who's very pregnant, you don't need to sign those people. If you see someone with eyes dilating, you don't need to sign those people. Do not create a risk situation for the company if you can."

When the HMOs weren't gaming the enrollment system, they were ensnaring doctors in a preapproval process that seemed, at best, ambivalent about patients' well-being. In one particularly egregious episode, recounted by George Anders in his book *Health Against Wealth*, two TennCare HMOs serving a rural community near Chattanooga refused to approve prescriptions for an antibiotic recommended in the treatment of a serious, but treatable, acute bacterial bowel infection. An epidemic spread, forcing town officials to close the schools and advise parents to boil children's clothing, until county public health officials finally prevailed on the HMOs to relent.

Critics of TennCare had predicted much of this. One symptom of the weak political support for Medicaid is that despite all the complaints about cost, the program never seemed to have enough money for what it was being asked to do. Many states skimped on administration and oversight, and this could help explain why the TennCare HMOs could get away with such irresponsible behavior so easily. And as states reached to cover as many people as possible, they sometimes reduced the payments they made for individual medical services to the point where many providers simply refused to see Medicaid patients, arguing that it would cost them too much money. (This was another reason so many Medicaid patients didn't get regular care.) From the program's first day, Medicaid had paid doctors and hospitals less than either Medicare or private insurance.

Medicaid, in other words, was already running on a bare-bones budget by 1993. Yet here was Tennessee asking HMOs to come in and provide the same services that the government had been providing directly, but to do it for even less money. The hope was that HMOs could pull off this feat because of their natural efficiencies—and that their inherent interest in promoting good health would lead to better patient care, not worse. But that hope was based on the idyllic notions of managed care derived from the old-style west coast group practices. And those idealistic organizations had long since ceded the managed care market to large commercial insurers who were accountable primarily to stockholders and had decades of experience at tweaking enrollment and benefits in order to increase profit margins.

This evolution was no different from the way HMOs had evolved in the private insurance market. But TennCare's enrollees were far more vulnerable than the typical middle-class American getting insurance through his or her job. The people enrolling in TennCare had less money, less education, and less experience with doctors, and—most significantly—they were in worse health. The profit-minded HMOs of the 1990s tended to be at their very worst when dealing with people who had multiple serious medical conditions, since those were the people who suffered most from arbitrary or overly aggressive restrictions on treatment. But these were precisely the people who naturally gravitated toward TennCare and whom, after all, Medicaid was supposed to be serving.

The same basic story line was unfolding across the country, as every state but Alaska eventually moved at least part of its Medicaid population into some form of managed care. Some of the most egregious stories came from Florida, where an investigation by a newspaper, the *Sun-Sentinel*, turned up stories of Medicaid insurers denying apparently necessary treatments. One HMO wouldn't authorize neurological tests for a ten-year-old boy who'd been badly injured in a car accident, on the theory that the tests weren't medically necessary. (The HMO relented after six months, when the boy's mother threatened to sue; the tests indeed discovered abnormalities.) Overall, the investigation found, twelve of the state's fifteen Medicaid HMOs were giving "poor" or "very poor" care to children and pregnant women, according to state audits. One possible reason was that many of the insurers were spending only half of their money on patient care, while their executives took home seven-figure salaries.

If the switch to managed care was the cautionary headline about Medicaid in the 1990s, the celebratory one was the ongoing expansion of Medicaid eligibility—which, combined with the establishment of a new federal health insurance initiative for the poor, extended health benefits to millions of people who hadn't had them before.

That initiative was called the State Children's Health Insurance Program, or S-CHIP, and became law in 1997 thanks to a rare show of bipartisanship during the second Clinton term. As with the original Medicaid law, nobody seemed to appreciate its significance at the time. But within a few years, it had enrolled some 3.95 million children and, in some cases, their families. By 2005, the proportion of children without health insurance was down to 12 percent—the lowest of any age group except the elderly. (Thanks to Medicare, virtually every senior citizen has basic health insurance.)

As with Medicaid, states ran S-CHIP with the help of federal money. And, for the most part, states were eager to do this—particularly since, by this time, they had money to spare. The states, benefiting from the strong economy of the late 1990s, suddenly found themselves overflowing with tax revenues. Even better, lawsuits against the tobacco industry, over its deceptive marketing practices, had won for the states tens of billions in settlement money, generally earmarked for health-related purposes. A few states with a strong progressive tradition, such as Vermont and Minnesota, used the newfound cash to enact dramatic expansions that achieved nearly universal coverage for their citizens. And with so much money sloshing around, even relatively conservative lawmakers found the program hard to resist. Among the states that would expand health insurance programs most aggressively was New Jersey, which had a Republican governor (Christine Todd Whitman) and a Republican legislature.

But the money would not last. The tobacco settlement was a one-time windfall, and state tax revenues began to fall in about 2001, once the economy started slipping. As before, the timing couldn't have been worse. The cost of health care was again increasing at its old rapid rate, while more people were losing their jobs and finding themselves in part-time or contracting jobs that offered no benefits. It was the early 1990s all over again, only this time the states couldn't turn to managed care as a magic bullet, since they were already using it.

It's not unusual for states to have fiscal problems in times like these, since most have balanced-budget amendments preventing them from borrowing their way through rough patches. This is why the federal government, which can (and does) run deficits, has sometimes stepped in to help during such periods. But by 2003, as state budgetary problems became acute, Washington was under new management: President Bush. And his contempt for government health insurance programs is a matter of public record.

During the 1990s, when Bush was still the governor of Texas, he had resisted calls to implement an S-CHIP program and then fought with the legislature to keep enrollment in the program down—even though Texas had the second-highest proportion of uninsured residents in the entire country. As a candidate for the presidency in 2000, he had criticized a plan by Al Gore to expand Medicaid and S-CHIP. And as president, he revived Reagan's old idea of a block grant for Medicaid, which Newt Gingrich had rediscovered and tried unsuccessfully to push through just a few years before. Bush's initiative for the block grant met the same fate as Reagan's had, and Congress eventually pushed through some additional assistance funding. But Bush was able to accomplish something unilaterally: he signaled to the states that he would give them more leeway to rearrange benefits or enrollment in ways that would, in the long run, reduce the role of government.

A few strongly conservative Republican governors seemed most enthusiastic about this possibility: in Florida, Jeb Bush, the president's brother, gave the state's Medicaid managed care companies more leeway to determine what benefits to cover; but there was little guarantee that the private insurers would provide the extensive services the

Medicaid population needed. (Adding to the problems was the fact that during the 1990s, some of the most infamous scandals involving fraudulent Medicaid insurers had taken place in Florida.) In Mississippi, Haley Barbour, the former chairman of the Republican Party, tried hard to cut 50,000 disabled and elderly people from the state's Medicaid rolls, backing down—partly—only after a public outcry persuaded the legislature to oppose him.

But it was a Democratic governor who would succeed in enacting the most sweeping cuts: Phil Bredesen of Tennessee. Bredesen knew a thing or two about health care: before getting into politics, he had made his fortune by founding a company called HealthAmerica—one of the first commercial HMOs to cash in on managed care during the late 1980s. This experience, plus his confidence in his own intellectual abilities (he had a physics degree from Harvard), convinced him that he could wring new efficiencies from Tennessee's Medicaid system, just as his HMO had generated financial savings—and hefty profits—in the private market.

A lot of what Bredesen proposed to do—such as reducing fraud by providers and recipients, and improving the use of information technology—made sense. But when those quick fixes didn't bring the TennCare budget under control, he unveiled a more straightforward plan: he would simply slash the program. More than 100,000 people who had qualified for TennCare because they were "medically needy" would lose their coverage altogether. Those allowed to remain in the program would have to make do with more limited benefits. The biggest change would be in the coverage of prescription drugs. "The sad reality is that we can't afford TennCare in its current form," Bredesen said. "It pains me to set this process in motion, but I won't let TennCare bankrupt our state. This is the option of last resort."

At the tiny North Terrace Family Clinic in Lawrenceburg, which serves both TennCare beneficiaries and people without insurance from across the Lawrence County area, the staff looked forward to Ernie's visits because his outsize presence seemed to lift everybody's spirits. He'd walk up and down the halls, ogling any leggy drug-company representatives who might be visiting and introducing himself to total strangers as if he were campaigning for office: "Ello," he would say in his thick accent, vigorously shaking hands. "I am Ernie Maldonado. It is nice to meet you." When the nurses wanted to get a rise out of Ernie, they'd mention something about his "Cuban" or "Mexican" heritage, sending him into a fit of rhetorical apoplexy. When he wanted to get at them, he would make jokes about needing a pregnancy test or an ultrasound test. "If it's twins," he said one day, grabbing his modest paunch, "I don't think I want to carry them." Of course, Ernie could have fun at his own expense, too. "Something is very wrong with me," he told a nurse one day, leaning into her and looking very grim. "I seem to be turning Puerto Rican."

Patsy Burks, the nurse practitioner who ran the clinic and was in charge of Ernie's care, was constantly amazed at his ability to project energy and enthusiasm—despite his litany of medical problems and the considerable discomfort they caused. And she thought the medications deserved a lot of credit for that. Although pharmaceuticals could not cure Ernie's ailments or eliminate the symptoms entirely, they could certainly reduce the pain. And that seemed to have a secondary effect on Ernie's mind-set. A few years earlier, as the conditions were piling up, Ernie had told Patsy he'd committed himself to making sure the diseases didn't get the best of him. Once the medications made the pain in his legs bearable, he began taking daily walks, which not only exposed him to fresh air but also kept fluid from accumulating in his knees and lower extremities. He became more social, too, picking up groceries and cooking for neighbors in the senior citizens' retirement building where he and Wanda had recently moved. He even found some new outlets for playful troublemaking: he became a regular caller on the local radio morning show, where the staff had nicknamed him "Dutch." (The host thought it sounded jolly, like Ernie himself.)

But when the state started talking about cuts in TennCare, Ernie's spirit subsided. He was supposed to be taking more than two dozen separate medications, the majority of them prescriptions. They were enough to fill a Jack Daniels whisky glass every morning, afternoon, and evening—the glass in which Wanda would prepare them for him. But under Bredesen's new plan, TennCare would pay for only five of these drugs—and only two could be name brand. (The rest had to be generic equivalents). Ernie frequently confided to Pat and her staff that he was worried, because now he wasn't the only one with serious medical problems. Like her husband, Wanda was a longtime smoker who now had severe breathing problems. (And, unlike Ernie, she hadn't quit.) She didn't require oxygen, but she did require two separate inhalers to open her lungs.

With just $500 a month on which to live, Ernie and Wanda figured they had no choice: they started choosing which medicines not to take. The over-the-counter vitamin supplements were the first to go, since they weren't essential; they'd just been prescribed so the two would be a little healthier. After that it was Ernie's diabetes pills. (Ernie figured he could live without them, since he could just take more insulin shots at night.) Ernie always wanted to cut his medications first, so that Wanda could have most of hers. But Wanda, who was in charge of buying and doling out the medicine, insisted that they take turns. "He would go without some of his medicines one month so I could get mine; sometimes I would do without mine so I could get his."

As the weeks wore on, Ernie got worse—not in a slow, steady decline, according to Pat, but in a series of rapid drops. He stopped roaming the halls on his visits to the clinic, because he couldn't walk more than a few steps without gasping for breath. He said that his back pain was getting worse, too. The clinic received a steady supply of samples from the pharmaceutical companies, and Pat did her best to make sure Ernie got the essentials—or whatever approximations she could find. "We would beg, borrow, and steal to get him the medicines, but he was on several we couldn't get," Pat later remembered. "Sometimes we would substitute one kind of inhaler for another kind that he was supposed to take. Sometimes anything is better than nothing." When the samples ran out—as inevitably happened, since it seemed as though half of the clinic's patients had lost TennCare and suddenly needed free medicines—Pat and another nurse would slip him some cash, so that he could buy enough medications to last him until more samples came in.

Ernie hadn't asked for the money, and when they offered it the first time he cried. "I was not brought up this way," he said. "It hurts my manhood." He apparently told nobody about the handouts, not even his wife.

Pat couldn't be certain that Ernie was suffering because he wasn't taking his medications. But Ernie wasn't the only one having a hard time. She saw some of her other patients struggling, too—and started complaining to the local papers. Then one day in October Jim Shulman, an adviser to Governor Bredesen, showed up at Pat's clinic, as part of a tour to assess the impact of the changes in TennCare and assure providers that help was on the way. Wanda and Ernie happened to be there that day, and Pat used them as an example of how deeply the cuts in the program were affecting the community. Shulman put his arm on Ernie's shoulder and told him the state was setting up a safety net program to deliver pharmaceuticals to the most seriously ill beneficiaries of TennCare, so that people like him could get the prescriptions they needed. "Hang in there," Shulman said. "In a few days we're going to have this all worked out."

Three days later, Ernie died.

In the weeks that followed, other stories of hardship poured in from around the state. Eventually, Bredesen and the state legislature found a way to restore coverage to at least some of those who'd lost it—apparently, this had been the plan all along. Cynics said that Bredesen did it for political reasons. By delivering pain early, he'd given himself an opportunity to provide relief—conveniently enough, just before he was up for reelection. Bredesen and his defenders replied that they were simply reacting to financial reality. They had made the cuts because Tennessee was running out of money to pay for its citizens' health care; and it was the money saved from those cuts, ultimately, that allowed the state to restore some benefits a year later.

Similar debates were taking place in state capitals across the country. But in Lawrenceburg, Wanda had more immediate thoughts on her mind: how to get along without Ernie. Her first night home, she wanted to sleep in the same bed he had used. "I'm going to bed on papa's bed," she told her children. They tried to talk her out of it, but she insisted. "When I laid down in that bed that night, it was like he put his arms around me. I felt it." She continued to sleep there afterward, eventually placing Ernie's ashes on an adjacent table, so he could be nearby. During the day, whenever she spoke about him, she broke into tears.

Still, Wanda also remembered how Ernie had nearly died once before, when he had his first heart attack, and what he told her about it afterward. He described seeing a field of grass, flecked with tiny flowers of the same size. He said he could see his own body, perfect, with all the blemishes gone. He said that he could hear Wanda calling him—"Honey, sweetheart"—and that he was trying to tell her, no, he wanted to keep going. But then he heard her scream, "Ernesto," and he stopped. Suddenly the grass and the field of flowers were gone. And he was back in his hospital bed, breathing through a tube. "If I ever start again," Ernie told Wanda after that night, "don't stop me."

So maybe, Wanda told herself, it was all for the best: "When he died, he had a smile on his face, a little smile like he was trying to think of something picky to say to me or something funny, that's the way he looked. And that's what keeps me going. I know he went at peace and not in pain. He didn't suffer. And I know he ain't suffering now."

SIX

CHICAGO

It was early one Saturday afternoon in the summer of 2001 when Marijon Binder found herself lying on a stretcher, staring up at the ceiling of the Stephens Convention Center just outside Chicago. Marijon, a live-in aide and companion to a handicapped elderly woman, was attending a trade show for people who work with the disabled. But Marijon, herself 62, had felt a pain in her chest just as she first stepped into the cavernous meeting hall. When she sat down on a nearby chair, in obvious discomfort, a security guard noticed her awkward movements and summoned medical help.

Marijon was a small, stout woman with a soft chuckle and owlish eyes that peered out from behind wire-rimmed glasses. But her humble appearance belied a stubborn personality, and, as the paramedics began examining her, she tried to brush off their assistance. The drive from her home in northwest Chicago had taken about an hour, during which she was breathing in warm, humid air through the open windows of her aging Dodge sedan. It was probably just the rush of artificially cooled air that made her lungs tighten up, she explained. She had felt this way before, there had been no lasting harm, and she would surely be fine now, too—at least until Monday, when she could see her own doctor. Besides, she had to get home to take care of her housemate.

But the paramedics, fearing that Marijon might be having a heart attack, practically insisted that she go to the hospital. And Marijon eventually relented, in no small part because of where they were proposing to take her: Resurrection Medical Center, the flagship institution of Chicago's Catholic hospital network. Marijon had no health insurance, and a Catholic hospital, she figured, could be counted on to show compassion—if not because of its traditional commitment to the poor, then because she just happened to be a former nun, who had served for thirty-five years.

When Marijon first arrived at the hospital, she assumed that her judgment was correct. As she tells the story, one of the first things she spotted was a set of three brown signs on the emergency room wall, explaining in three languages (English, Spanish, and Polish) that Resurrection would take care of any patient regardless of ability to pay. When the hospital staff gave her the standard consent forms to sign, Marijon crossed out the sections stating that she would be responsible for paying all medical charges, writing in large block letters "I have no insurance." She also tried waving off the medical help again, but with no more success than she'd had at the trade show. Her doctors assured her that there was nothing to worry about—that the hospital had a way of taking care of such charges—and got back to monitoring her heart. Eventually, a hospital administrator showed up with an application for financial assistance.

Marijon would end up spending two days at the hospital. She wasn't having a heart attack, but her tests indicated she'd had one previously and the medical staff wanted to observe her. She finally got the doctors to approve a release, but only by promising to see her own doctor, with whom she had a prior arrangement for free care, the following Monday. By this time, Marijon says, she had already completed the financial aid application, except for some material she would later mail from home. She left, confident both that she was healthy and that the hospital would give her a break financially.

But Resurrection, which would later say it never received a completed charity application, hadn't written off the bill. Instead, it sent Marijon an itemized invoice, listing some 60 different charges, from $3 for a single dose of Tylenol to $609.25 for some laboratory work. The total for her two-night stay came to $11,000.

What happened next is a matter of some dispute. Marijon says she called Resurrection, explaining her situation, and was told by an employee of the billing department to do nothing and wait—that the financial assistance office was probably just very slow in processing her application. Resurrection says it has no record of that phone call. In any event, the bills didn't stop coming. And Marijon—at this point peeved that the hospital was still trying to charge for services that, after all, she had never really wanted in the first place—decided not to call back. She handled her housemate's medical bills all the time, and confusion like this was not uncommon. Eventually, the bureaucrats would sort it all out.

They did, only not in the way that Marijon had expected. About a year after the hospital visit, on a warm September night, an officer from the Cook County sheriff's department showed up at the bungalow-style home Marijon shared with her housemate. The sight of the policeman on the front porch was such a surprise that it took Marijon a moment to notice the piece of paper in his outstretched hand. It was a summons with Marijon's name on it. Resurrection was suing.

One reason that Americans have embraced, or at least tolerated, a health insurance system that leaves tens of millions without adequate coverage is the widespread belief that a medical safety net exists to take care of those people. And the most important part of that safety net has always been the hospitals that provide charity care—public hospitals owned by the government; and private nonprofit hospitals, most of them academic or faith-based. For most of the twentieth century, the public hospitals had primary responsibility for charity care, with academic and religious hospitals handling the overflow. But by the end of the century, skepticism about big government coupled with shrinking municipal and state budgets had led local authorities in such major cities as Philadelphia and San Diego to downsize, sell, or close their public hospitals.

This shifted more responsibility for charity care to private, non-profit hospitals—a burden they were seemingly well-suited to handle. Many of the institutions, particularly religious hospitals with names such as "Daughters of Charity," cultivated an image of themselves as charitable organizations that put a greater premium on doing good than on making money. And all nonprofits benefited from tax exemptions, worth millions or even tens of millions of dollars to each institution—tax exemptions that, according to law and tradition, the hospitals are supposed to earn by serving their communities.

But the actual history of nonprofit hospitals, as opposed to their mythology, suggested that they might not be so enthusiastic about inheriting ultimate responsibility for the uninsured. Throughout the twentieth century, many nonprofit hospitals saw caring for the economically vulnerable as a burden rather than a calling. Even those hospitals most committed to charity work worried constantly about their financial viability, carefully circumscribing their care to keep it within their perception of what was financially feasible. And starting in the late 1980s, just as an explosion in the number of uninsured Americans was increasing the demand for charity care, these hospitals were facing new economic pressures of their own—thanks to efforts by both government and the private insurance companies to contain the rising cost of medical care.

The turnabout frightened the leaders of the hospital industry, who warned darkly of impending bankruptcies and cuts in services. It also worried advocates for the uninsured, who understood the role non-profit hospitals played in providing access to care. But the dominant voices in policy making for health care, including those in academia, business, and politics, hailed this unleashing of market forces. As they saw it, competition and the threat of financial failure would force hospitals to become more efficient. A few might not survive the shakeout, but the rest would emerge stronger, ultimately producing better care for more people—and for less money.

In many ways that is precisely what happened. A few financially weak or poorly run institutions closed, and the rest adapted by managing their operations with a more explicitly businesslike sensibility. Hospitals streamlined supply chains, hired professional money managers, and improved information systems—in effect, reengineering themselves in the way countless other industries had been doing for years. But the quest for efficiency also had a harsher side. As hospital administrators became ever more determined to eliminate waste and strengthen their institutions financially, they inevitably began to scrutinize another drain on funds, treating the uninsured. The conflict between money and mission, always present at these hospitals, intensified. And oftentimes, money won.

The hospitals of Resurrection Health Care trace their lineage back to the late nineteenth century and the early twentieth—a time when advances in medical science and waves of European immigrants conspired to make modern hospitals ubiquitous in the urban landscape.

In the late 1800s American cities were teeming with newcomers who had come to run the factories, tanneries, and stockyards of the industrial revolution. As these masses from places such as Killarney and Kraków established enclaves in large urban centers, calls went back to Europe for religious orders that could serve each group's specific language and cultural needs.

Health care was high on the list because it was difficult to treat people if a language barrier prevented them from explaining their symptoms. And among the various faiths, Catholics were the ones who seemed to answer the call most enthusiastically, most likely because of the enormous importance they placed on administering last rites to the dying, preferably in words the dying could understand. (This need seemed all the more important in a Protestant nation where anti-immigrant and anti-Catholic sentiment was known to run high.)

It was just such considerations that brought both the Holy Sisters of Nazareth and the Sisters of the Resurrection, the two sponsoring orders that would eventually merge their hospitals into Resurrection Health Care, from Poland to Chicago in the late 1800s. Although the Sisters of the Resurrection would not actually establish their hospital, the future Resurrection Medical Center, until the 1950s, they purchased the land it would occupy in the 1920s, while it was still an undeveloped marsh on Chicago's far northwest side. According to legend, Sister Anne Streselecka, head of the order, "looked out across . . . this awful swampland and said there's going to be a hospital there someday." The Holy Sisters of Nazareth, meanwhile, had already established Saint Mary of Nazareth Hospital; in 1895, they had set themselves up in a three-story building nestled under the el tracks, on the edge of the city's west side where most of the new immigrants lived.

That location assured the hospital a steady stream of patients with economic needs to match their medical problems. A Polish meatpacker in the city's famous stockyards might have to work sixty hours a week, often standing knee-deep in water, just to pay for his family's room and board. Disfiguring industrial accidents, whether from a stray butcher's knife or a careening railroad car, were daily occurrences. While the middle class lived in newly built housing on the outskirts of the city, enjoying such amenities as electricity and steam heat, immigrant families crowded together into filthy one- and two-bedroom tenements that bred tuberculosis, diphtheria, and cholera. They couldn't always afford their own rent, let alone a night in the hospital. In 1897, a survey of patients at yet another Catholic hospital destined to become part of the Resurrection network—Saint Elizabeth's, founded by a German order—showed that 40 percent of the patients were charity cases. According to an authorized history of the facility, "no one was turned away for lack of financial means."

That may well have been the case. As a rule, though, nonprofit hospitals were not always eager to take on the indigent. Like their nonsectarian counterparts, religious hospitals of the time frequently had no compunction about distinguishing between the deserving and undeserving poor, usually leaving the latter to the likes of Cook County Hospital and other publicly owned facilities. They also looked for ways to maximize revenue, by promoting their medical technology, courting paying patients, and promising private rooms away from the crowded wards of the charity cases. In fact, religious hospitals tended to have even higher percentages of paying patients than nonreligious hospitals.

This early effort at "cost-shifting"—the term economists use to describe the way hospitals use paying patients to subsidize care for the poor—allowed hospitals to survive the early twentieth century. In fact, several of Chicago's hospitals were able to expand during this period, thanks to generous donations. And as health insurance evolved, hospitals developed cost-shifting into an art. Particularly in the early years after Medicare, hospitals were flush with cash—a situation that, at least in theory, gave them the financial resources to cover the costs of people who still had no insurance coverage.

But that economic environment disappeared once both private employers and the government became determined to control their rising health insurance bills. Among other things, the government—as part of its effort to control Medicare spending—repealed a provision of the law that had allowed hospitals to pass along their capital improvement costs to the program. That provision had effectively given the hospitals a blank check with which to finance expansions and purchases of new technology. Hospitals had been (and remain) notoriously dependent on capital investment because health care technology changes so quickly; a hospital without the latest scanner or the equipment to perform the latest surgeries will quickly lose patients to a hospital that has such devices. Without government to back these purchases, nonprofit hospitals would have to get their money by issuing bonds. And that meant satisfying Wall Street's investment analysts, whose exclusive interest was the hospitals' bottom line.

Suddenly nonprofit hospitals had to compete for a shrinking pool of reimbursements and investment money. And they weren't competing just with each other. One other consequence of the creation of Medicare in the 1970s was the growth of the for-profit hospital industry, as entrepreneurs jumped at the opportunity to claim part of the Medicare largesse. They established three major conglomerates— the Hospital Corporation of America, Humana, and Tenet—all of which instantly became darlings of Wall Street, attracting enormous amounts of capital. And since—unlike the nonprofits—they were bound by neither law nor tradition to serve the public interest, they had more leeway to keep their profits high (and their investors happy) by maximizing profitable services, cutting out unprofitable services, and skimping on costly care for the poor.

Among Chicago's religious hospitals, the adjustments to this newly competitive market were swift and sweeping. The most extreme reaction came from Michael Reese, a Jewish hospital on the South Side with a reputation for academic excellence rivaled only by its reputation for generosity to the uninsured. But as it lost its more affluent patients to the suburbs, the money from billing (and from donations) dried up. In the 1980s, administrators at Michael Reese became so panicked by their bottom line that they actually converted the hospital to a for-profit entity and sold it to Humana. (Years later, Humana would sell it to Columbia/HCA.) Once the hospital had overstocked Popsicles so that young patients could always have them; the new managers eliminated such frivolities. As a former executive ruefully conceded later, "You can't run a $200 million operation on nostalgia."

No other hospital in Chicago underwent such a dramatic legal transformation, in part because non-profit hospitals were always particularly dominant in Illinois. Instead, hospitals improved their competitive position by consolidating. During the 1990s, several of the city's Lutheran and Evangelical hospitals merged, eventually forming the Advocate Health Care network, one of the largest such hospital systems in the entire country. Shortly thereafter the city's Catholic hospitals, which had also begun merging several years previously, came together under the banner of Resurrection.

Consolidation allowed hospitals to take advantage of economies of scale—for example, by reducing their paid staff as they eliminated services that were now duplicated within their network. More important, though, the mergers gave the hospitals much greater clout in bargaining with insurance companies. By the early 1990s managed care had the ultimate leverage when the time came to set reimbursement rates. If an insurer was not satisfied with the prices a hospital was offering, it could simply walk away from the negotiation, confident that few patients would mind losing access to just one hospital. But the situation changed dramatically once groups of hospitals, rather than individual institutions, appeared across the bargaining table. Now if an insurer walked out on negotiations, it would be depriving its beneficiaries of access to many more hospitals. Thus the insurance companies would have to sweat it out, worrying if their beneficiaries would flee to other plans. (A vivid illustration of this changing balance of power came in 2003, when Advocate Health Care threatened to walk away from negotiations with Illinois Blue Cross-Blue Shield, the state's largest insurer. The two sides ultimately worked out a deal; although details were never disclosed, insiders in the financial community said that Advocate succeeded in driving up the reimbursement rates substantially.)

Another conspicuous transformation in Chicago's nonprofit hospitals took place within the boardrooms. It was fine to have a nun or a minister in charge of one small hospital. But as knowledge of managed care contracts became more important than familiarity with the gospel, religious hospitals began hiring professional hospital administrators, many with experience in the for-profit world. By the late 1990s, both Advocate and Resurrection had professional administrators at the helm—and paid them accordingly. In fiscal year 2002, Resurrection's CEO, Joseph Toomey, received $2.3 million, including benefits and onetime bonuses. That same year, five executives at Advocate received more than $1 million each.

Catholic leaders insisted that they were not embracing commercialism so much as providing an alternative to it. Cardinal Joseph Bernadin of Chicago had long advocated consolidation of the local Catholic hospitals as a bulwark against further encroachments by for-profit enterprises like Columbia/HCA. In a speech to the Harvard Business School Club of Chicago in 1995, he noted that the purpose of the Catholic hospital was to heal patients and "not to carn a profit or a return on capital for shareholders." And at Resurrection, officials felt they had fulfilled that mandate, bristling at any insinuation that they had now become a "chain." Said one spokeswoman: "We are a health care ministry, a health care system, and certainly not a 'chain' any more than the churches in a specific city are a 'chain' of the Catholic Church."

In many respects, that was true. Nuns were no longer in charge of Resurrection, but they remained on the board and in several lower executive positions. They also gave regular orientation meetings to new employees about Resurrection's mission, stressing that "our mission and values are not separate from how we do our business."

But in countless ways, big and small, Resurrection was now clearly a leaner, more conspicuously commercial enterprise. The tiny hospitals founded and run by eastern European nuns a century before had become a massive, multimillion-dollar conglomerate that dominated the health care market in Chicago's northwestern neighborhoods. And while Resurrection's corporate structure still had a nonprofit organization at the top, down below were for-profit subsidiaries, including a captive insurance company based in the Cayman Islands. The main entrance at Resurrection Medical Center had the mission statement inscribed on the wall, but the wall itself was a slab of marble that ran from the floor up to the sixteen-foot ceiling—part of an ornate lobby that would look perfectly at home in any four-star hotel. Patients seeking to feed their souls could stroll into the interior courtyard and pray before a statue of Saint Joseph, the "model of workers, protector of families." Thirty or so feet away, those seeking comfort of a more sensory nature could sidle up to Resurrection's coffee bar, which offered lattes, mochas, and cappuccinos supplied by Seattle's Best Coffee.

The story of Marijon Binder's childhood reads a little like something out of a novel by Steinbeck. She'd been born during the Great Depression, one of five children of a French-Canadian mother and German father living near Buffalo, New York. The family lived in welfare housing without much heat; and, one day while she was five, a kettle of boiling water spilled on her back while she and her siblings were trying to warm themselves at an open oven. The resulting burns kept her out of school for nearly a year, forcing her to repeat kindergarten.

Her parents had a rocky marriage—in part because her father, who worked in a brewery, was an alcoholic—and decided to try for a fresh start when Marijon was about seven, packing their earthly possessions into an old Plymouth sedan and driving to California. The car kept breaking down, until one night, when the family had reached a small desert town near Palm Springs, they decided to take shelter in an abandoned farmhouse. Marijon says that nobody ever asked them to leave, and they would remain there for several years, raising rabbits and chickens while Marijon's father worked in construction and at a gas station. Later, they moved into a trailer and then into another abandoned house—this one on the property of an elderly heiress (apparently related to the Hearst family) who agreed to let them stay if Marijon would read to her every day after school.

Marijon hated all the moving around; she was continually falling behind in school and had difficulty meeting new friends. But she loved and admired her parents, particularly her father—a kind man who, despite his drinking problem and the family's tough economic circumstances, always found ways to make the best of their situation. And perhaps because both of her parents were devout Catholics, who seemed to draw strength from their deep faith, Marijon decided from an early age not only that she wanted to help other people but that the best way to do it was through the church, the one constant in her otherwise tumultuous young life. At thirteen, she began reading books about various orders, and at seventeen, she joined one.

In her particular religious community, nuns could choose to pursue nursing or teaching. Marijon chose the latter, teaching her first class of eighth-graders when she herself was just eighteen years old. Several years later, after completing both her education (earning a bachelor's degree in Greek and Latin) and her formal religious training, she took her vows for life and moved to San Francisco, where she taught geography and social studies. She also managed to do some professional writing, including a teacher's edition of a geography textbook; that earned her an invitation to Chicago, to work for a publishing company there. It was supposed to be a six-month assignment, but once she had arrived, she never left. She eventually founded her own non-profit organization for training teachers about geography and world cultures, a project her religious order supported by setting her up in a house on Chicago's North Side.

The house was in a working-class neighborhood with many elderly residents. And it wasn't long before Marijon came to know many of them—and their troubles—as they battled the infirmities of age. Marijon began helping them, taking a particular interest in their pets. She had loved animals since her days at the farm in California, and now she began taking in neighborhood strays. Eventually, she started her own organization dedicated to helping the elderly keep their pets, so they wouldn't have to have the animals euthanized. She also became particularly close to one of her volunteers, an old woman named Eleanor with no living family whose own health was deteriorating.

When it became clear Eleanor couldn't live on her own, Marijon took her in. Not long afterward, Marijon's superior told her to return to California. Marijon thought that Eleanor and the rest of the neighborhood needed her. And, more important, she believed that God wanted her to stay. She refused the order, prompting the superior to begin dismissal proceedings. The whole process took two years. Marijon went through formal hearings and even retained a lawyer— but when the procedure was done, the ties were formally severed. Although heartbroken, Binder says, she was confident she had done the right thing by staying close to the people who needed her most.

———

Eleanor was still in Marijon's care in 2002, when Marijon got the subpoena from Resurrection Health Care. And when Marijon realized she'd have to appear in court, on a specific date and time, she says she immediately began to worry about what to do with Eleanor. "That was a huge imposition, because whenever I leave my [house-mate], I have to pay somebody $15 an hour to stay with her. . . . We just didn't have enough money to be doing that." She called an attorney friend, who walked her through the legal process, and then she contacted a nun from her former order, who suggested that she go to Resurrection's management directly. It was a Catholic hospital, after all, the nun reminded her. Maybe the management would help.

Marijon called and arranged an appointment with an official from the Resurrection "mission" office. But the meeting had lasted only a few minutes when it became apparent that Marijon wasn't going to be getting any relief. Resurrection did have a small fund to finance care for nuns. But it was only for current nuns. Since Marijon was no longer affiliated with her order, she didn't qualify. Not one to give up easily, Marijon explained her saga. Never mind her religious service. Couldn't she get help simply because her small income made her a deserving candidate for charity care? Again, the answer was no. It was too late to consider such options, the official said, after consulting with a higher-up. Now that the case was in the hands of Resurrection's lawyers, Marijon would have to deal with them.

On October 2, Marijon did just that, composing a letter in longhand to the attorneys at Grabowski and Greene, the firm that handled many of Resurrection's debt collection cases. In the letter, which she sent by fax, Marijon said all of the same things she had said in the hospital's mission office. She said that she was a former nun who had "devoted the last ten years to caring for the elderly" and was now working full-time for one infirm woman. Then Marijon painted a detailed picture of her financial situation. "I receive no salary nor income from any other source," Marijon wrote, explaining that the two housemates lived off the elderly woman's Social Security and pension, which in a typical month left no more than $40 after expenses. Sometimes, Marijon pointed out, the two even shared "Meals on Wheels" in order to save money. Just in case the attorney doubted her word, Marijon made sure to attach the two documents she'd brought with her to the meeting at Resurrection: a mailing from the bank (showing an end-of-month balance of $41.27) plus an affidavit from the Chicago Housing Authority vouching for her financial status. Then, after recounting her past efforts to secure financial assistance, she asked one more time for help in applying for charity care.

Resurrection's attorneys were no more accommodating than the mission office had been. She got no response to the fax itself. When she finally got somebody on the phone, she says, the lawyers at Grabowski told her that if she wouldn't make minimum payments of at least $100 a month—which Marijon said she couldn't afford, because even her extremely frugal lifestyle left her with just $40 in a typical month—then she'd have to go to court.

It isn't every day that former Catholic nuns are sued by Catholic hospitals. But apart from that idiosyncrasy, stories like Marijon's were actually quite common. Between 1999 and 2003, Resurrection sued hundreds of former patients over outstanding debts. According to court records, at least seventy of these people were "indigent" by the standards of Illinois—the court had certified that their incomes were less than 125 percent of the poverty line. And eventually these lawsuits would start getting attention, thanks in part to a local labor union that was investigating Resurrection's business practices. The local, a chapter of the American Federation of State, County, and Municipal Employees (AFSCME), was in the midst of a campaign to organize hospital workers. In 2003, it released a report called "Debt Collection Practices at Resurrection Health Care: An Appeal for Compassion," featuring stories of patients sued by the hospital—most of them seemingly hapless victims of a corporate hospital bureaucracy run amok.

Such stories weren't limited to Resurrection Health Care. Hospitals all around Chicago—and, indeed, all across the country—would soon be knee-deep in controversy over their billing and collection, as a result of muckraking by entrepreneurial reporters and activism by other unions (which, like the AFSCME local in Chicago, were trying to organize hospital workers.) Among the most infamous stories were some from Yale-New Haven Hospital in Connecticut. In one case, memorably described in the *Wall Street Journal*, the hospital forced a man to spend two decades paying off the $16,000 debt from his late wife's hospital care, eventually putting a lien on his house and wiping out his life's savings. One reason it took so long for the man to pay the bill: Yale-New Haven was charging 10 percent interest.

The Service Employees International Union (SEIU), which had been trying to organize Yale hospital's workers, had a hand in publicizing these tales. But labor unions apparently had nothing to do with the revelations about Provena Covenant Medical Center, a Catholic hospital in Urbana, Illinois. There, it was community activists who brought the practices to light. At Provena Covenant, attorneys would ask judges to issue "body attachment" orders, empowering the police to haul debtors into court, by force if necessary. This practice was considered so severe that, according to another story in the *Wall Street Journal*, even the giant retail operators like Sears Roebuck and the Ford Motor Credit Company (which handles credit for Ford) had previously prohibited their collection agencies from using it. Not Provena Covenant. In fact, during one such episode, police officers handcuffed a forty-year-old diabetic man in front of his wife and son in order to bring him to a hearing—all for an outstanding hospital bill of $579.

Reports of these tactics sparked widespread outrage: "To put so much silent agony on hapless, hardworking, low-income Americans—that's just absolutely unacceptable as conduct," said Uwe Reinhardt, a widely respected health economist at Princeton. And, perhaps because broader solutions for the uninsured seemed out of reach, elected officials soon seized on hospitals' behavior as a political cause. By early 2004, everybody from Chicago aldermen to Washington power brokers were calling to revoke the tax-exempt status of hospitals that used aggressive collection methods against the uninsured—a threat that became a reality for Provena Covenant, which lost its tax-exempt status. Attorneys were getting involved, too, rounding up patients who had been sued by the hospitals to see if they wanted to counter-sue, alleging, among other things, that the hospitals had violated the compact to provide community service. Eventually more than 400 hospitals would be named in such class-action lawsuits. And, ominously for the hospitals, the leader of the litigators filing most of these suits was Richard Scruggs, the attorney from Mississippi whose pursuit of the tobacco industry during the 1990s had famously yielded a $206 million settlement.

━━━━━

Faced with the prospect of such huge financial losses, either through litigation or through changes in tax status, the hospitals fought back, defending their collection practices as both fair and necessary. Indeed, Resurrection made no apologies for its firm treatment of Marijon or other uninsured patients, calling the stories in the union's report misleading and challenging accounts that former patients were giving. The hospital insisted that it made financial assistance available to those who needed help and who were willing to follow its procedures for applying. The problem, Resurrection said, was that many of the people who wanted assistance didn't really need it—and that even those who did need it often refused to cooperate when the hospital tried to help.

Hospitals have long approached potential charity cases with such wariness. In 1935, an essay by a hospital superintendent in New Jersey described how his facility would send social workers to the homes of patients who asked for free care, just to make sure they were truly indigent. In one instance, a social worker filed a report stating that a patient seeking charity care "lived in a beautiful house. Door was answered by the maid who refused to disturb her mistress, as she was still sleeping. Apartment beautifully furnished. Telephone. Family of two. Rent $50 a month." That patient was denied further access to the hospital.

Resurrection could provide no similar instances of patients who were seeking free care while secretly living in modest luxury, but it did provide enough details to suggest that the stories were, at the very least, more complicated than they seemed at first. Many patients who ended up getting sued by Resurrection had, at one time or another, failed to complete necessary paperwork, failed to apply for available state assistance, or failed to respond when the hospital attempted to reach them. Even Marijon Binder, who had taken the initiative in asking for financial assistance from the outset, would later say that she had probably erred by letting so many bills come to her house without an answer. A spokeswoman for Resurrection concluded: "It is our policy to be as compassionate as possible to patients, but in order for us to respond to their financial situation, we must ask them to share facts relating to their income and other factors that pertain to the ability (or inability) to pay. . . . All we ask for is cooperation."

But the fact that Marijon, who had a master's degree and plenty of experience dealing with health care bills, had so much trouble with the charity care application process suggested that perhaps "cooperation" wasn't nearly as easy as Resurrection made it sound, particularly for uninsured patients with little education and poor English. And plenty of critics wondered whether this wasn't perhaps intentional, part of a broader strategy to limit its care of the uninsured. Federal law prohibited hospitals from literally turning away patients with emergencies just because those patients didn't have insurance. It also required hospitals to post large signs advertising this fact, which is why all hospitals have the large notices that Marijon saw when she first came to Resurrection. But that federal law said nothing about whether hospitals could then pursue payment from such patients—nor did it require hospitals to notify uninsured patients about possible financial assistance. As late as 2004, the Resurrection Medical Center E.R. had no similarly prominent posters advertising its charity care options—let alone giving instructions on how to apply for them.

And then there were the handouts from a training session for Resurrection's financial counselors. One handout, part of a Microsoft PowerPoint presentation, enumerated the documents patients had to submit in order to be considered for charity care: a driver's license or other valid proof of local residence, at least four paycheck stubs for the previous two months, the most recent federal income tax return, all current bank statements, copies of all W-2 or 1099 forms from the previous year, and the two most recent rent or mortgage receipts. Another handout then instructed counselors on how to process them—"After you receive the returned application: Check to see if the application has been filled out in its entirety. Check to see if patient and/or co-applicant have signed the reverse side. Check to see if patient has provided all documents which are required to process the application. If any of the above are *missing/incomplete*, the application should be 'DENIED'!"

Officials at Resurrection insisted that the hospital network was not making a concerted effort to reduce its care for the underserved. In fact, many of the physicians who worked there seemed dumbstruck by the mere suggestion: "We've never been told to cut back," said one. "In fact, we're always being told to be more mission-oriented and finding patients that need care." To back up their claim, they cited anecdotes, such as one about a homeless man whom paramedics transported to a Resurrection ER in 2002. Over the ensuing months, doctors performed several surgeries on him, sometimes admitting him to the intensive care unit—probably the single most expensive part of the hospital. Eventually Resurrection's social workers tracked down the man's family—in Poland. Once he was stable enough to travel, the hospital staff arranged for him to return to his family. Resurrection estimates that in this case it absorbed $140,000 in hospitalization costs alone.

Overall, however, the hospital really did seem to be doling out less financial assistance than before. From 2002 to 2003, according to statistics filed with the state, charity care at Resurrection's hospitals fell by one-third. Resurrection blamed this apparent decline on an accounting quirk, but critics noted that the drop was most precipitous at Saint Elizabeth and Saint Mary of Nazareth hospitals, the two facilities within its network surrounded by the heaviest concentrations of poverty and non-English-speakers.

Certainly, such a reduction would have been entirely consistent with what numerous studies suggested was happening around the country. As reimbursements to hospitals declined and managed care pursued cost containment, hospitals reduced their charity care. As one study warned in 1997, "continued expansion of managed care throughout the country, particularly in light of the decline of insurance coverage, may undermine the ability of hospitals to continue to provide uncompensated care."

In Chicago, proceedings over medical debt take place at the Richard Daley Center, an intimidating thirty-one-floor building of brown steel that straddles almost an entire city block near a corner of the downtown Loop. Marijon began her first day there in typical fashion—waiting in a long line outside the clerk's office, where defendants must register. But the wait turned out to be a blessing of sorts. Defendants in debt cases must normally pay a court appearance fee of $90 for debts under $5,000; or $140 for debts over $5,000. Just as Marijon was about to have her turn, she overheard somebody else in line talking about the possibility of qualifying for indigent status, which exempted defendants from paying the fees. Marijon made sure to ask about that when her turn came up—and succeeded in getting her fees waived.

Marijon also asked about getting legal assistance. The good news was that a local law school staffed a help desk. The bad news was that it was open only on Tuesdays and Wednesdays—and this happened to be a Monday. So she made arrangements for another day of professional help for her housemate, arriving a few minutes before nine the following morning because she knew that the help was available on a first-come, first-served basis. To her chagrin, three people had already signed up ahead of her.

It would be several hours before she finally gained an audience. Her frustration mounted as the attorney told her that she stood little chance of actually winning her case—that the best she could probably do was stall the proceedings, particularly since the judge who would be hearing her case had a reputation for siding with the hospitals. But the attorney was also pleased to have a client so eager to fight the case, particularly one so articulate and, better yet, with the moral credibility of a former nun. So he walked her through the process that lay ahead of her, explaining how to file briefs for discovery, and so on. He also tipped her off to the hospital's largest legal vulnerability: its prices.

Marijon had thought the prices on the medical bill seemed outrageous, but she also figured they were just a reflection of how expensive medical care had become. What she hadn't known until now was that hospitals like Resurrection routinely discounted care for everybody *except* the uninsured, as a response—once again—to the recent upheavals in their financial environment. Before managed care, hospitals limited themselves to one set of charges—say, $1,000 for use of a surgical room. But when private insurers began demanding lower prices, hospitals would typically meet those demands by offering the insurers special discounts—so an insurer with a 50 percent discount would pay only $500 for the surgical room. Inevitably, as the hospitals sought to generate more revenue, they would raise the "sticker price" for services, which, just as inevitably, prompted the insurers to demand even steeper discounts the next time both sides negotiated. The gap between charges and insurance payments grew; and by 2003, one authoritative study showed that charges could be four times higher than what insurance companies actually pay when their beneficiaries get treated. This penalized the uninsured, who had nobody negotiating on their behalf. Surveys by AFSCME would later find that the charges passed along to uninsured patients at Resurrection ran higher than what insurance companies paid for the same services.

Marijon filed a brief demanding that the hospital justify its pricing policy. She also asked the hospital to produce the original consent form (the one in which she'd written that she had no insurance) and to produce witnesses who were in the emergency room while she was being treated. But Resurrection's attorney did not cooperate. Instead, he filed his own briefs dismissing most of the requests as "not relevant." The briefs went back and forth for about six months that stretched through Chicago's dismal winter, requiring several appearances before judges and, accordingly, several more trips downtown—each one a three-hour round trip on public transportation. (Marijon worried that parking downtown would be expensive and difficult to find.)

April finally came and, with it, a final hearing before the judge. By now, Marijon was all too familiar with the scene at actual courtrooms upstairs, which was hardly the stuff of *Law and Order*. There were no juries and, except in rare cases, no defense lawyers, either—just a judge, some court personnel, and four or five collection attorneys scurrying about, dispatching cases of debtors who never answered their summonses. On those occasions when a defendant was present, judges customarily asked him or her to meet with the collection attorney privately and work out a mutually acceptable arrangement. Every day, the halls were full of people huddled in conversations, as the attorneys peppered bewildered-looking defendants with questions about their assets, employment, and access to cash.

Marijon, though, hadn't come this far to cut a deal. As the hearing began, the judge asked whether Resurrection had produced the witnesses Marijon had requested. No, the attorney explained—he'd simply brought one staffer who could explain hospital policies. The judge cut the attorney off, saying he was interested only in people who had seen Marijon and could speak about her experience in the emergency room. Then he listened to Marijon's story, thought about it, and gave a ruling on the spot: he was permanently absolving her of responsibility for the bills. Resurrection's attorney reacted by "hopping up and down," in the words of one eyewitness, telling the judge that the ruling was illegal and that he would appeal the decision.

━━━━

By late 2004, as the talk of political and legal repercussions grew louder, many hospitals had announced they would change their billing and collection practices—whether by making more people eligible

for financial assistance or by imposing new restrictions on debt collection activities. Provena Covenant created a board of community
representatives that reviewed all debt cases slated for referral to attorneys. (It also appealed the revocation of its tax-exempt status.)
Advocate Health Care announced that henceforth it would pay attorneys a fixed fee if they decided to refer cases back to the hospital
for consideration for charity rather than pursuing collections through
the courts or payment plans. Resurrection made two conspicuous
changes, revoking the geographic restrictions on charity care and
posting in its emergency room waiting areas large colorful signs
about charity care.

Of all the defenses the hospitals made during the controversy, the
one that was probably hardest to deny was that they were becoming
scapegoats for a much larger problem. It was entirely possible that the
critics were right—that the hospitals could afford to be more generous and that, under tax exemption laws, they were obligated to be.
Indeed, work by Nancy Kane at the Harvard School of Public Health,
one of the most respected experts on hospital finance, had suggested
that hospitals cutting back on charity care were frequently using it as
an excuse—that the better, and proper, response to cost pressures was
simply to manage themselves better. But, Kane noted, it was absurd to
suppose that hospitals, even with better management, had the capacity to deal with the present uninsured population—let alone one that
would be larger in a few years. Indeed, she said, if government took
the money flowing to nonprofit hospitals through tax exemptions and
redirected it instead into insurance for all the uninsured, it would buy
only one month's worth of coverage.

As Marijon Binder recounted her saga a year after the fact, she
confessed to similarly mixed feelings. "The whole experience was very
demeaning. It made me feel very guilty; it made me feel like a crimi-
nal. Every time I came from downtown I felt like I was coming from
jail and doing something wrong. I can imagine what it does to other

people who are poor." And yet, sitting on the same front porch where she'd first gotten that subpoena, she couldn't bring herself to blame the hospitals entirely. She believed in Catholic hospitals. And while she thought Resurrection's representatives could have been more considerate, she also understood the pressure they were under. "What really needs to be changed is the whole health care situation; we need to get everybody insured. Making private hospitals go bankrupt isn't the answer, either."

SEVEN

LOS ANGELES

The drive from West 190th Street to Vernon Boulevard is a straight shot up the Harbor Freeway, right through the heart of South Central Los Angeles. Although it can take forty-five minutes at rush hour, the traffic has usually thinned out by one o'clock in the morning—the time when a security guard named Jose Antonio Montenegro used to commute home from his job at an electronics factory. It was an easy drive he had come to know well and, on clear nights when he could glimpse the twinkling lights of the skyscrapers downtown, it was even vaguely picturesque.

But on a July evening in 2005, Tony, as he called himself, never saw the lights of downtown. In fact, the only lights he could see were the red and white ones on the backs of cars in front of him—and even they were a blur. Tony had started feeling woozy and disoriented a few minutes before, while he was finishing his shift. By the time he had pulled his pickup truck onto the highway, his vision started to go, too. He thought about pulling over, but even for somebody who lived in a relatively rough part of town, this was no place to stop in the middle of the night. So he pressed on, slowing his car to well below the speed limit, relying on his knowledge of the route to guide him home.

Somehow he made it, parking his truck on the street and stumbling his way into bed, not bothering to wake his wife or young son—hoping, apparently, that with a good night's sleep he'd feel right again. When he woke up the next morning, however, his vision had deteriorated even further. The blur was gone; now he could see nothing but darkness. Tony's wife, Gloria, promptly took him to a local clinic, where it didn't take a doctor long to deduce what had happened. For years Tony had battled diabetes. Now he had suffered one of the well-known complications of the condition: he'd had a stroke.

Tony was still relatively young at the time, just fifty, with a thick, muscular build. That made him a relatively unlikely victim of stroke, even with the diabetes, just so long as he got the regular medical care the condition requires. But regular medical care is exactly what Tony had lacked for most of his adult life. Although he'd been working full-time almost continuously since arriving in the United States from El Salvador nearly twenty years before, he'd had health insurance only sporadically. On the income he and his wife made, medical care had become a luxury item—and luxuries were not something they could generally afford. So Tony did not get his regular checkups or keep track of his blood sugar or take the medications that had been prescribed for him. Medically, he was a walking time bomb.

Tony's situation was not unusual. In fact, it was—and is—entirely typical for its time and place. The urban poor face more health hazards than most Americans, for such reasons as the toxic dust of slum housing and the destructive effects of drug abuse. Yet the urban poor also have less access to medical care—because the jobs available to them frequently don't provide decent insurance and because the safety net facilities on which they depend simply aren't up to the task. Although this is hardly a new problem, it has gotten demonstrably worse as the strains on private and public health insurance have grown in the last twenty to thirty years. And nowhere is this more true than in Los Angeles, which by the end of the 1990s had come to resemble health care's version of the apocalypse. It had the nation's single largest concentration of people without insurance—nearly 2 million, or about one out of every three adults. It also had what was, arguably, the single most overmatched network of publicly financed clinics and hospitals.

For the uninsured and underinsured of the inner city in Los Angeles, the lack of adequate medical care was just one of many routine hardships—not so different, really, from the difficulty of finding a job that would pay a decent wage, a car that would start in the morning, or a neighborhood where kids could play safely on the sidewalks. And in a perverse sense, this baseline of everyday struggle spared them the financial crises that had, over the years, affected middle-class families like the Rotzlers, the Hilsabecks, and the Sampsons. If you lived in an economically depressed part of Los Angeles, or any other major American city, you probably didn't own a house to lose and hadn't accumulated savings to deplete once illness struck. Medical bills couldn't ruin you financially, simply because, to be ruined, you had to have amassed some wealth in the first place.

Still, even in the bleakest corners of South Central Los Angeles, some families clung to one precious asset: hope. It was the hope that they could break out of poverty and, eventually, put their children on a path to a better, more secure life—the same hope that had first lured Tony and Gloria Montenegro to the United States many years before. But for them and for millions of other families in America's sprawling ghettos, hope required health. And when the latter deteriorated, so did the former.

Like much of America, Los Angeles has undergone a dramatic economic transformation in the last 25 to 30 years. For most of the postwar period, the city was a center of industrial might—a hub of manufacturing activity as prosperous and vital to the country's well-being as Detroit or Pittsburgh. But instead of turning out cars or steel, Los Angeles made airplanes. Rockwell, Lockheed, McDonnell Douglas—all of them had major operations in the city, where they produced not just glimmering flying machines but an affluent existence for the working and middle classes.

The decline of the cold war changed all that. Cuts in national defense spending devastated the aerospace industry—and, with it, the area's base of stable, decent-paying jobs. Although manufacturing in Los Angeles didn't completely die, it changed forever, shifting more to nonunionized industries such as garment making and upholstery, where the work paid far less. Meanwhile, the largest sources of new jobs were the service industries, which also had less generous wages and benefits. For a while, the number of Americans moving out of southern California actually exceeded the number coming in. But the state's population continued to balloon anyway, because of a constant influx of foreigners, predominantly Latino and Asian. Along with the area's African-American community, the immigrants provided the bulk of the new low-wage workforce.

Tony was part of this wave. As a child growing up in Santa Ana, a medium-size city in the north of El Salvador, he had worked on construction projects with his father, with whom he lived. (His father and mother had separated.) But from an early age Tony had aspirations for a professional life; he studied hard in grade school and was accepted by a university, where he had started to learn accounting. Tony says he was a good student—that is, until the political unrest began. It was the late 1970s, and El Salvador was descending into a civil war between the American-backed government and Marxist guerrillas. Although most of the actual fighting took place in the countryside, violence seeped into the cities as well. With the economy grinding to a halt and the university stopping classes, Tony decided to get out—

partly because it was the only way to get ahead and partly, he admits, because he was young and leaving seemed like an adventure. With a group of friends, he made his way to Mexico and then, driving in an old Ford Mustang, over the border to Texas. Another two days of driving brought him to Los Angeles, where some fellow refugees from El Salvador introduced him to a man who owned a garment factory and was looking for workers.

The factory was typical of the new economy in Los Angeles. It was an early twentieth-century cinder-block building that had been subdivided into large rooms scattered over ten floors, each room filled with immigrants from Asia and Latin America—most of them without official papers—who made clothing, shoes, or upholstery. (Tony got his working papers a few years later and eventually became a full U.S. citizen.) With only windows and fans to cool many parts of the building, it was literally a sweatshop. But for workers like Tony, it was also a home. By day, he sat hunched over a sewing machine, stitching together fabric covers for couches and being paid a few dollars for each new piece he produced. By night, he slept in a living area he and some of the other workers shared in the cavernous basement, two floors below ground level. There, the men cooked with hot plates and toasters, using stray cinder blocks to mark off their personal space, which they shared with the occasional mouse. Just one bathroom was available to them, a few stories up in a part of the factory that made shoes. Otherwise, the only plumbing consisted of a water hose snaked in from outside and a huge, empty oil tank underneath that the workers used as a drain and gradually filled over time.

Tony was OK with the living conditions. In fact, he says, he was happy to have the job—and a base from which to explore his new world. Like that of many other recently transplanted foreigners, his English was still very poor. But even when he was living in El Salvador, Tony had developed a passion for American and British music, from Sinatra to the Rolling Stones. One night, a man who was renting

some space for himself in the factory—one of the few Anglos there—
played an old album that Tony recognized. Tony used the opportunity
to introduce himself by pointing at the record player and using one of
the few English words he knew: "Beatles." He and the man, named
Richard Berghendahl, became fast friends.

Eventually, the management changed and the new managers decided to renovate the building. Fortunately for Tony, he'd recently begun dating Gloria, who was also an immigrant from El Salvador and
whom he'd first met at a party. Gloria had come from the countryside, fleeing the violence there. (She would later tells stories of waking
up and walking out her front door to find dead soldiers lying at her
feet.) Tony liked the fact that she was serious and mature, the type of
woman who talked openly about her plans for the future—somebody,
in other words, with whom he could imagine spending his life. So
the two soon married, in a storefront marriage parlor, and moved together into a house they found on the northern edge of South Central,
just a mile or so from the Coliseum sports arena. It was a relatively
small place—a stand-alone dwelling behind a larger home, with just
two bedrooms, a kitchen, and a bathroom. But with Tony between
jobs and Gloria making not a lot of money doing domestic work at a
rest home, the $350 monthly rent was still more than they could af-
ford on their own. They offered to split the house with Richard, who
gladly accepted (since he had been living at the factory, too, and the
alternative was sleeping in his old Volkswagen van).

It was 1986 when Tony and Gloria moved there, near the peak of
the nation's crack cocaine epidemic that had turned Los Angeles—
like most American cities—into a virtual war zone. The Montenegros
were on the front lines, because across the street from their next-door
neighbor's home was a liquor store with an outdoor pay phone. This
phone was a magnet for drug dealers, and violence was common. One
sunny afternoon, according to Tony, a crowd approached a young
man standing in front of the liquor store. A minute later, the crowd

had dispersed and the man was lying splayed on the sidewalk, blood streaming from a fresh bullet wound in his head.

At the time, the Montenegros' neighborhood, like most of south-central, was predominantly African-American. One day in 1992, while driving home from an English class he was taking, Tony heard that verdicts had been issued in the case of Rodney King, exonerating the policemen who had been accused of beating King—a young African-American man—during an arrest a year before. By the time Tony turned onto Martin Luther King Boulevard, the main thoroughfare that ran closest to his house, unrest over the verdict had already broken out. When he got to his cross street, he saw rioters torching the gas station on the corner and looters ransacking a nearby grocery store. Another group had broken into the liquor store across the street from Tony's home, so he quickly went inside and took Gloria up to the second floor. There, the two spent the night listening to rifles firing into the air and glass crashing onto the sidewalk, while the glow of their burning city lit up the sky.

Tony says he would have liked to move elsewhere—to Culver City or La Brea or anywhere that was a little less dicey. But, like his dream of resuming his business studies, that hope would have to remain on hold, given the limited income on which he and Gloria lived. After a stint working for a commercial printer, Tony had begun working as a security guard. The jobs paid modestly, at or just above the minimum wage. And although Tony had insurance while he worked for the printer, he didn't when he moved to security work. The first company that employed him didn't offer coverage. The second apparently did, but Tony says he didn't find out about it until after he'd been working there for several weeks—past the official enrollment period during which new workers were allowed to apply for coverage.

To this day, Tony says he's not certain whether or not the company informed him of the option. Given his still limited English, it's possible that he was told and simply didn't understand; it's also possible that nobody bothered to inform him. (Such confusion is common in immigrant communities.) But it wasn't something about which he had thought to inquire, since the previous job didn't offer benefits and he didn't expect them. Nor was it something he pondered particularly hard. After all, he'd always been healthy anyway.

———

Every year, the World Health Organization (WHO) publishes information from countries around the globe, tabulating statistics such as the number of children who receive all their recommended vaccinations and the number of people who die from various types of cancer. Every year, the Department of Public Health of Los Angeles County goes through a similar exercise, compiling statistics about the county's eight geographical districts. The Montenegros lived on the edge of what was called Service Planning Area 6 (SPA-6), which stretched from downtown Los Angeles in the north all the way to the city of Compton in the south. And by global standards, this area didn't stack up particularly well. The death rate from cervical cancer in the Montenegros' district was more than twice the average for the rest of the United States. It was also higher than the average in most of the developed world and even in medical backwaters like rural China. Another set of data showed that the children of families in the Montenegros' section of Los Angeles were far more likely than their counterparts in the industrialized world to be born at dangerously low weights—and, afterward, to develop asthma or diabetes during childhood. Going down the list of measures reveals the same story over and over again. According to virtually all meaningful statistics, the health of people living in and around South Central was substandard—and in a few instances, it bordered on what many would consider third-world conditions.

Service Planning Area 6 included South Central, which had become predominantly Latino; and Watts, which remained largely African-American. The two communities had distinct health profiles, for reasons of both genetics and culture. (Among other things, Latinos overall seemed to benefit from a traditional diet rich in legumes.) But at least in this part of Los Angeles, the two groups did share one characteristic that helped explain the overall poor health outcomes: poverty. More than one-third of the residents had incomes below the federal poverty level of around $20,000 in yearly income for a family of four. And many more residents had incomes hovering just above that level. Amid these economic conditions, drug use, unprotected sex, and other forms of destructive behavior were common—which was why, for example, the infection rate for AIDS in South Central Los Angeles was climbing even as the nation as a whole was bringing the epidemic under control.

The link between behavior and poor health undoubtedly reinforced public antipathy toward government spending on health programs in the inner city; if the people living in and around the Montenegros were shooting themselves up with drugs or shooting each other with bullets, just how much money was everybody else supposed to spend on bringing them back to health? But destructive conduct had a way of visiting consequences on people who had nothing to do with it, such as toddlers growing up in homes filled with secondhand smoke or teenagers catching stray bullets in drive-by shootings. And sometimes the people who chose to engage in unhealthy habits did so for legitimate reasons. Eating a proper diet wasn't easy in neighborhoods that lacked large grocery stores with affordable fresh produce. (Not surprisingly, consumption of fruits and vegetables in SPA-6 was the lowest in Los Angeles.) And in particularly blighted sections of town, parents who returned from work late had to weigh the benefits of outdoor exercise for their children against the risk of violence. At one point, a housekeeper at a local hotel who lived about a mile and a half from the Montenegros explained that she kept her children inside, watching television, lest they get caught in cross fire while playing outside. She wasn't paranoid: two weeks before the interview, a shoot-out outside her window had crippled a teenager.

In any event, behavior was not exclusively responsible for the health problems of the inner city. Far from it. Living in South Central Los Angeles meant living in close proximity to the region's smog-producing freeways and soot-belching factories. All too frequently, it also meant living in substandard housing with old, chipped paint and decaying insulation. Community activists repeatedly claimed that these environmental hazards were part of the reason for the region's high rates of mortality from cancer. And while studies generally weren't able to substantiate the link with cancer, they did tie environmental hazards to at least two other problems that were reaching epidemic proportions, particularly among children: lead poisoning and asthma. (On the opposite side of the Harbor Freeway from the Montenegros' house, a neighborhood adjacent to a group of metals factories had acquired the nickname Asthma Town.) At Saint John's Well Child Center, one of several clinics that provided discounted and free pediatric service to the community, the majority of children who received care had dangerously high blood levels of lead, most likely from deteriorating house paint. Doctors there had also become accustomed to dealing with less harmful, but no less disturbing, medical issues arising from substandard housing—such as treating rat bites and removing cockroaches from children's ears.

The other critical way economic hardship affected the health of people living in South Central Los Angeles was by cutting them off from routine medical care. Nearly half of the working-age adults in the area had no health insurance at all—a situation that reflected not only a lack of private, job-based coverage but also a confusing array of public insurance programs that supposedly existed to fill in the gaps. Together, the county, state, and federal government ran what were really a dozen separate health insurance plans, each one targeting a narrowly defined group of low-income people. There was Medi-Cal, California's version of Medicaid, which covered children, pregnant women, adults, families, and those who needed long-term care—but with different income thresholds for each group. Healthy Families, which was California's version of S-CHIP, also had varying income guidelines and its own registration process. The county ran its own insurance programs, based on ability to pay, but only for people who didn't qualify for the other programs.

Residents who were technically eligible for these programs frequently didn't know about them—or couldn't figure how to apply. The programs also required continual renewals, requiring recipients to reaffirm their low-income status with extensive documentation; as a result, people frequently complained that they had been unexpectedly thrown off the rolls, sometimes because they'd failed to submit paperwork a second time and sometimes, they said, for no good reason at all. The stringent requirements for enrolling in these programs ostensibly reflected a desire by the county and state to weed out people trying to defraud the system—a legitimate concern, to be sure. (And, of course, not all those who told a journalist that they had filed paperwork on time had actually done so.) On the other hand, it was an open secret that officials in California—and many other states—used these requirements as a way of trimming program rolls without attracting negative attention in the media.

All this went a long way toward explaining why one-fourth of the adults in South Central Los Angeles had no regular source of medical care—a category that included Tony Montenegro. Although Gloria had gone onto Medi-Cal when she was pregnant, and although the program covered their son, Antonio, when he was born, Tony as an employed man was not eligible for these programs. So he simply didn't bother with medical care. In fact, he can't remember going to the doctor even a single time in the early 1990s—not until one day when he noticed that his toe was starting to hurt.

As it turned red and swollen, it became too painful for walking and Tony decided he would have to get it checked out. He knew of a local storefront clinic, one of many in the area that provided cheap, à la carte medical care to immigrants on a cash business. For a $25 office visit fee, a doctor examined the toe, took a history, and decided that Tony should get tested for diabetes. Tony had a hard time believing that this could be the problem; except for the toe, he says, he felt absolutely fine, with no fatigue, no dryness of the mouth, none of the classic symptoms. But he got the test anyway and, indeed, it came back positive.

Now Tony had a serious problem. He was supposed to test his blood sugar daily, more often if he was under stress or had eaten badly. He was also supposed to take pills to control the sugar and—if those didn't work—inject insulin. But Tony figured that if he followed those orders to the letter, the test strips plus the medications would cost a few hundred dollars a month. And that was not money he and Gloria had lying around. Gloria was doing domestic work at a rest home for the elderly, making pretty much what Tony did: close to the minimum wage. That put their household income at around $25,000—and their expenses had increased since Antonio came into their lives. In addition to the extra food and children's supplies, the couple had decided to move into the main house on the same lot, a slightly larger apartment that had a living room. Between rent, food, utilities, clothing, gas, and the minimum required insurance for the car, they didn't have much at the end of the month.

So sometimes Tony got his supplies and medicine. And sometimes he didn't. He hoped for the best and, for a while, he seemed to be getting by—until that night on the freeway when he had the stroke.

The Los Angeles County-University of Southern California Medical Center may be the most recognizable hospital in the United States, if not the entire world. Since 1973, its exterior has appeared in the opening credits of the ABC television show *General Hospital*. From a distance, the edifice still looks more or less as it does on television, its familiar white pillars and darkened windows rising high atop a hill in the Boyle Heights neighborhood, just east of downtown. But up close, the exterior details hint at a purpose more serious—and an existence more gritty—than the stuff of daytime soap operas.

On the facade, above the hospital's main entrance is a sculptural group: the angel of mercy taking care of a woman with a newborn child on one side, and a poor suffering man on the left. (The three people symbolize the three stages of life.) Just below, within the arched entryway, is an inscription pledging that the staff will provide medical services "without charge" to the infirm and the poor, "in order that no citizen of the county shall be deprived of health for lack of such care and services." The sculptures, the inscription, and indeed the whole massive structure are monuments to the aspirations of the early 1930s, a time of grand public works premised on the notion that government could solve common problems such as disease and economic deprivation. During that period, virtually every big city in America built and operated its own version of County Hospital. And although many of those cities would eventually sell or close their facilities, on the theory that private hospitals were better suited to the job, Los Angeles never did. During the 1990s, County-USC Hospital continued to operate as before, serving as the provider of last resort for the millions of poor who called Los Angeles home.

But in sixty years the burden had grown enormously, because there were a great many more people to serve—and the hospital, whose inpatient capacity had gradually shrunk from 2,000 to 750 beds, had no place to put them. The most visible impact was in the hospital's emergency room, which was perpetually filled to capacity, and by the late 1990s was on "ambulance diversion" three-fourths of the time. In the waiting rooms, flies buzzed around while people sat listlessly in chairs, sometimes bleeding and moaning as they waited long hours—or even days—to be seen.

One night, not long before Tony had his stroke, Gloria Montenegro was one of those patients. She'd been experiencing a stomachache for several days, hoping it would just go away. But when the pain became intense, she feared that this might be a real emergency, and she went to the only hospital she knew would take her: County. As Gloria tells it, a member of the staff—Gloria doesn't remember whether it was a doctor, a nurse, or somebody else—asked her about her condition and instructed her to sit tight, offering some Maalox to settle her stomach for the time being. Then Gloria waited. And waited. Ten hours passed, and as Gloria looked at the woman sitting next to her—who appeared to be bleeding from a miscarriage—she decided that it might be another ten hours, or more, before the hospital finally got around to her. She had Tony take her back home, and she eventually recovered; but she still got a bill for nearly $1000. After persuading the hospital to reduce it by a few hundred dollars, she ended up paying it over several months.

As bad as the conditions were in the waiting room, they were worse inside the emergency treatment area, which was littered with patients lying on gurneys, just waiting for space in the wards above. At one point, according to one of the emergency department's former directors, the average waiting time for actual admission from the emergency room to the hospital was about twenty-four hours. (The unofficial record was seven days.) Privacy basically ceased to exist amid these conditions, as sounds and smells circulated freely among the fifty or sixty patients packed into an area built to hold only forty.

And there was a clinical impact, too. In 2003, a group of doctors frustrated by the situation claimed that they knew of at least four patients who had died because of long waits for beds at County. Although the subsequent state investigation did not substantiate those charges—the patients, according to the report, would have died even if they had been seen more promptly—it concluded that overcrowding and the resulting delays in admission were compromising the quality of care and had almost certainly led to worse medical outcomes. Among the cases the report cited: A newly diagnosed tuberculosis patient who needed admission and isolation (25 hour wait for an inpatient bed); an HIV sufferer with fever and altered mental status (66 hours); a psychiatric case who'd apparently tried to commit suicide by overdose (32.5 hours); and somebody with aortic dissection, an emergency condition that sometimes requires surgery (45 hours).

Nor were the problems confined to patients coming through the ER. Medhat Elsadani was a fifty-five-year-old restaurant manager from Downey who in 2003 suffered a heart attack and was brought initially to California Hospital, downtown. Once it became apparent that he'd need cardiac catheterization, a procedure California Hospital couldn't perform, the staff decided to move him to County-USC—the only technologically suitable hospital that would accept him as a transfer, since he had no insurance. But according to Elsadani's account, County-USC had no open beds in its ICU—and it wouldn't have any for seven more days, while Elsadani waited in considerable discomfort. Eventually a bed opened up and Elsadani made the move. But it would be another three days before he could actually get the procedure, during which time the staff on several occasions told him that it was time for the surgery but then canceled the surgery because a more pressing emergency had bumped him down the list.

The surgery itself involved two stages: first determining the extent of blockage in Elsadani's arteries, and then using a metal wire to clear it. When the surgery began, the idea was to do these two stages in rapid succession. But after the first stage, when the doctors discovered 100 percent blockage in one artery, they had to hold off on the second—again, because another patient with a more pressing emergency needed care. Elsadani was shunted to the ICU, with an open incision, until the doctors could return to his case. Recovery took place in a crowded, noisy room that Elsadani shared with six patients. "At one point, one patient got out of his bed and started walking across the room with blood gushing everywhere," Elsadani later recalled. "His street clothes were soaked in blood and he had not been changed. The bathrooms were extremely dirty, with signs of vomit, urine, and feces. . . . The conditions in this room were horrific and I could not wait to get out."

Elsadani gave that account in 2004, as part of a lawsuit against the Los Angeles County Board of Supervisors, the seven-member board that runs the county government, over a controversial overhaul it had recently endorsed. By that time, the county health system included five major hospitals, including County-USC, plus a series of health centers and smaller clinics that were supposed to provide routine and outpatient care. For at least a decade, the system had teetered on the edge of insolvency, as more and more people came to depend on it. Twice the Clinton administration had rescued it with large financial bailouts, the first of which Clinton announced in dramatic fashion during a press conference on the tarmac of the Los Angeles International Airport. But now a financial crisis loomed again. And the Bush administration—which had neither the political inclination to help a Democratic stronghold nor the philosophical commitment to throw large sums of money at large government bureaucracies—had already stated that another bailout would not be forthcoming.

The county had to do something, so it decided to push ahead with a reengineering of the health system that it had actually begun, in fits and starts, years before as a condition for receiving financial assistance from the Clinton administration. The plan included simple management reforms, which were desperately needed in a system that was, by all accounts, run poorly at every level. It also involved closing the trauma center at Martin Luther King-Drew Hospital, one of the five main public hospitals with a well-earned reputation for shoddy care. But the county board was also proposing some more dubious changes—in particular, taking County-USC Hospital down from 750 to 650 beds and selling or closing altogether the Rancho Los Amigos rehabilitation hospital.

The county didn't try too hard to sugarcoat the plan: "If anything," said one official, "we would like to be adding more health services. It's a matter of not being able to afford to do it." Still, there was hope that improving the system's finances might free up more money for primary care clinics, thereby forestalling health problems before they became serious and, as a result, reducing the burden on hospitals. But the county had tried that once before, as part of its effort to meet the conditions of the second bailout by Clinton. And it didn't have the predicted effect. Rather than reduce the need for hospital services, as hoped, providing more primary care to the uninsured had actually increased the demand for these services. This was probably because giving more examinations to more of the uninsured had turned up many medically serious conditions that otherwise might have gone untreated. Cutting down the number of hospital beds again, this time more dramatically, promised to make the crowding even worse and, in the words of an independent consultant who assessed the plan, "lead to preventable patient deaths due to prolonged waiting times."

In the end, a federal judge blocked the proposed closings, citing arguments like these, and the county managed to extract a smaller, one-time infusion of cash from the Bush administration. But even the plantiffs, an alliance of community activists and unions, understood that this was a qualified victory. The county system was clearly in crisis, partly because it was mismanaged but mostly because the job of providing for the uninsured in Los Angeles had simply become too big for it to handle—at least with the resources at its disposal.

Tony's experiences with the county system were considerably better than Gloria's had been, probably because he was treated largely through a nonprofit clinic that operated under a special public-private partnership begun in the late 1990s, in response to Clinton's bailouts. It was called the Clinica Monseñor Oscar A. Romero, named for an archbishop and human rights advocate who had been gunned down in the early days of El Salvador's civil war. The name, and the clinic's location in the heavily Latino Pico Union neighborhood north of South Central and west of downtown, made it a magnet for Latin American immigrants—particularly those, like Tony, who because of a sudden medical crisis found themselves in need of health care and with no clear sense of where to find it.

Clinica Romero, like the Saint John's Well Child Center and a few other clinics around the city, did truly heroic work—providing not just medical care but a whole array of social services designed to meet the particular needs of the population they served. Many of the clinics had innovative day care programs that combined regular pediatric checkups with reading instruction; some also offered dental care, filling a need few outsiders even seemed to realize existed. (Intense tooth pain was a leading cause of absences from school; and, of course, missing school made it that much harder for children growing up in low-income neighborhoods to get ahead.) Most of the clinics had mental health programs of one sort or another, often featuring addiction treatment. Some, like Saint John's, sent outreach workers into homes to help parents inspect for lead paint—and then petition either their landlords or the housing authorities for remediation.

It was, according to Tony, the staff at Romero that had first made it possible for him to take proper care of his diabetes, by providing him with discounted or free supplies and medicine. And it was the staff at Romero who, upon diagnosing the stroke, arranged for him to get specialty care that he could afford. They helped Tony enroll in the county's insurance plan, for which he qualified, and they arranged for him to have eye surgery through a physician who operated at nearby Good Samaritan Hospital. Although Tony had to wait two months for the surgery, it worked, restoring some eyesight to one eye.

Clinica Romero was a federally qualified health center, one of hundreds around the country. These clinics, which represented one of the most successful—if least known—health initiatives from Washington, received federal money in exchange for a promise to see uninsured patients free or at considerable discounts. The program dated back to Lyndon Johnson's Great Society. But more recently, at least, its biggest champion may have actually been President Bush, who, although always skeptical about large government insurance programs, had long talked up the importance of local charity clinics—perhaps because they had a well-documented record of success in his home state, Texas. Under the Bush administration, funding for the federal clinics had increased steadily—enough to establish as many as a few hundred new clinics.

But the rising support for clinics came at a time of falling support for other safety-net programs, at least relative to the exploding need for them. In 2005, a group of experts carefully examined all the different ways that the federal government paid for health care services that benefited the poor—whether through Medicare, Medicaid, or direct subsidies to clinics—and how that spending had changed in the first years of Bush's presidency, when the rising number of uninsured Americans made the demand for services so high. The unambiguous result was that spending had not kept up with the new need. And, really, this is what had been happening in Los Angeles all along. Strictly in terms of dollar figures, more and more money kept going into helping the uninsured of Los Angeles County get medical care. But given how expensive medical care itself was getting, and how many more people kept losing insurance and looking to the safety net for help, it just wasn't enough.

In Tony's case, the care he got also wasn't enough—but mostly because it was simply too late. His doctor at Clinica Romero could keep him from getting another stroke. But neither she nor the specialists to whom she referred him could fully undo the damage already done. He could see out of only one eye, and then with a limited range of vision. Before the stroke, Tony had been an avid reader, sometimes consuming books—usually thrillers or mysteries—for two hours a night. After the stroke, he all but stopped reading, because just getting through a page became a struggle. The vision problems also interfered with Tony's ability to enjoy fatherhood. When he tried playing catch with Antonio, who was now ten, he'd miss the ball completely or let it hit him square in the face, because he simply couldn't see it.

And, perhaps most difficult of all, the stroke had forced Tony to stop working—at least for a while. With his limited vision, objects off to the sides were a blur, and objects right in front of him were nearly invisible. Once, when he dropped his glasses on the kitchen table, he could not find them until Richard, his neighbor, came by and pointed them out to him; they had been inches in front of him. Tony applied for disability coverage, but it took more than six months for the state to process his application and verify his status, during which time he had to sell the family car to help pay the rent. And even after the disability checks started coming, the financial struggles continued, since the disability benefits paid only two-thirds of Tony's former salary. He was not eligible for full replacement because his vision, though severely limited, was enough for him to do some work.

Sitting in his house one afternoon in 2006, underneath a picture of the Last Supper that his wife had hung on the wall, Tony agreed he'd probably have to find a new job. But he wasn't sure where. He thought about working in a store and minding a cash register, but feared he couldn't keep up. "Everything is complicated; in another job you need to be fast," he explained, saying that he had spells of confusion and disorientation. "And if somebody drops something, they drop their coins, I won't know where they are." Eventually, after a second eye operation restored more of his vision, he decided instead that he'd look for more security work. Even if he couldn't move around as quickly as before, it was something he at least knew how to do. He'd long since given up his dream of taking night classes in business, or even moving into a better neighborhood. At this point, just climbing back to where he was—getting by and surviving—would be satisfaction enough.

EIGHT

DENVER

Russ Doren loved his wife dearly. That is why he was so desperate to keep her from coming home.

It was August 1990, nearly three months after Gina Doren had first been admitted to the psychiatric ward of Porter Hospital in Denver. She was no stranger to the facility, or to mental illness. For most of her adult life she'd battled severe depression—a result, apparently, of sexual abuse she had suffered as a child. And since 1981, the year she gave birth to her son Kory, she had been in and out of the hospital more than a dozen times. But this particular admission was shaping up as one of the scariest. Gina had tried to slit her wrists twice, first using broken glass from a lightbulb she'd taken out of a ceiling fixture, and then using shards of a soda can she'd taken from the trash. Russ had discovered that second effort during visiting hours, when he found her on her bed with blood on her arms. For a while after that, the hospital staff wouldn't even let Gina go into her room alone, because they feared she'd take the opportunity to hurt herself again.

And yet Russ had hope. Hospitalization had pulled her back from the brink before. Even if it took a while, once the doctors figured out the right combination of medication and once she'd gotten enough intensive therapy, she was capable of resuming normal activities—as a student at a local community college, a wife to Russ, and a mother to Kory. In those intervals, she was able to overcome (or at least suppress) her psychological demons. And for a while, anyway, life for Russ resembled what he had imagined it would be when they first married more than a decade before. They had wonderful times together, Russ would remind himself, and they could still have more—if only Gina had a little more time to get better.

But whether Gina would get to stay at Porter Hospital was no longer simply a medical issue. It had become a financial question, too. And it was not one the Dorens could answer to the hospital's satisfaction. Russ, a high school biology teacher, had generous medical coverage for him and Gina through the affluent suburban district where he worked. But his policy, like most private insurance policies, provided only limited benefits for the treatment of mental illness. In contrast to the coverage for physical problems, the insurance entitled them to only forty-five days a year of psychiatric inpatient care. The coverage included a 50 percent co-payment, too. That meant the Dorens would be responsible for half of all the hospital bills—with Porter charging nearly $1,000 a day while Gina was a patient there.

A few weeks earlier, when the outstanding balance had approached $50,000, Porter's business office had begun asking Russ when he planned to start making payments. He had responded by saying that money was on the way—literally, that the check was in the mail. But Russ had no money to send. The cost of Gina's psychiatric treatments over the years had long since depleted the Dorens' savings. And their efforts to secure additional assistance from either the school district or the state had failed. The staff at Porter had apparently concluded as much; that was why an attendant had intercepted Russ while he was on his way to Gina's room for another visit. As Russ recalls, the hospital's representative explained that Gina couldn't stay any longer, because of the mounting debt and the Dorens' inability to make payments on it. "We're a business," Russ says he was told, right before being urged to enroll Gina in a day program, where she would be safe while Russ was at work and where she could continue her rehabilitation.

Russ was not optimistic about that option. From past experience, he knew that spaces in day programs were hard to come by—and, even when space was available, these programs usually required payment up front. But mostly, Russ was worried about Gina, whose psychiatric troubles he'd been helping her to battle for nearly as long as he had known her. Russ could not recall Gina ever coming home in such poor condition. That's why he had lied about the bills and fought so hard to keep her in the hospital—and why he was more than a little scared about what might happen next.

Russ's situation was common in 1991. And it would be even more so today. In fact, all the forces pushing health care out of reach for Americans—the decay of employer-sponsored insurance, the spread of managed care, the commercialization of private charity institutions, the overloading of strained public programs—have affected people with psychiatric disease, and their loved ones, most directly. If the world of health care can be divided into "haves" and "have-nots," people with mental illness are a hugely disproportionate share of the "have nots."

But the story of what has happened to people battling psychiatric disease is a warning even to those Americans who will never have to deal with it. The mentally ill are not merely the ultimate example of how American health care can fail its citizens at their moment of greatest need. They are also a harbinger of the kind of world we will create if health insurance, both private and public, continues to marginalize people with ailments that are the most difficult and expensive to treat. It will be a world full of hardship that leads to tragic, and perhaps avoidable, endings—like the one to which Russ and Gina were hurtling.

———

Care for people with psychiatric disease has always lagged behind care for people with physical ailments, going at least as far back as the 17th century, when Europeans routinely tortured, burned, or hanged the mentally ill as witches. The American colonists were not quite so harsh. But even in this country, "treating" the mentally ill consisted largely of confining them, preferably in almshouses or jails that segregated them from mainstream society. And that is how things were more or less until the 1850s, when Dorothea Dix, a Unitarian schoolteacher and social reformer, began her crusade on behalf of society's "unfortunates." Part journalist and part agitator, she would tour institutions filled with the insane, write up descriptions of their wretched conditions, then read these essays aloud on the steps of state capitols. (As a woman, she wasn't allowed to enter these capitols.) Dix wanted the states to assume responsibility for the mentally ill by establishing asylums—places where they could live in relative comfort and, just maybe, get better. And she succeeded: vast, publicly funded institutions to house the mentally ill soon sprang up across the country.

This was the first time American society had ever approached psychiatric illness with such an explicitly humanitarian impulse. But the results soon proved disappointing. Decades after Dix's crusade, conditions in U.S. mental institutions had become nearly as deplorable as those in the almshouses Dix had exposed. States were using asylums to warehouse criminals, debtors, and the homeless. Grossly underfunded and understaffed, these hospitals were typically better at holding and hiding the mentally ill than at treating them.

One obstacle to obtaining better resources and oversight for these institutions was the stigma attached to mental illness. Another was the lack of sophisticated scientific knowledge about mental disease, which led to ambiguity over the definitions of illness and uncertainty over the effectiveness of treatments. That started to change in the early twentieth century, with the spread of Sigmund Freud's teachings. The two world wars further increased awareness of mental illness. Just as the rejection of draftees because of physical problems helped promote the campaign for universal health care in the late 1940s, the development of acute psychiatric problems among former combat soldiers strengthened the nation's commitment to helping, rather than shunning, people with mental illness. Thanks in no small part to a boom in the construction of psychiatric hospitals during the 1930s, under the auspices of President Roosevelt's Works Progress Administration, the number of Americans admitted to hospitals for mental illness climbed past 500,000 in the 1950s.

This surge in hospitalization eventually became controversial, as researchers began to suggest that many patients would be better off in outpatient or part-time settings, based in their own communities, where—with appropriate treatment and supports—they could reintegrate with mainstream society. A wave of deinstitutionalization ensued, and to this day the rate of hospitalization for mental illness remains far below the peak it reached in the 1950s. (In 1955, three-fourths of all psychiatric treatment took place in hospitals. By 1985, the trend had reversed, with three-fourths of all treatment taking place outside the hospital.) But even as society was pushing the mentally ill out of hospitals, it was demonstrating an ever greater commitment to treating mental illness. Calling for a "great national effort" to help the mentally ill, President Kennedy substantially increased funding for research and training while signing into law the Community Health Centers Act of 1963, which put up the first serious money to build outpatient psychiatric clinics.

One indicator of the changing attitude toward mental illness was the evolution of health insurance benefits. The early Blue Cross plans of the 1930s hadn't covered psychiatric illness at all, leaving that responsibility to the state institutions that had been bearing that burden for most of the previous century. This situation had started to change by the late 1950s, when comprehensive health insurance had spread to most of the working population, but the majority of employer-sponsored plans still excluded psychiatric benefits. The lingering stigma had something to do with this, but so did a fear—on the part of private insurers and those who paid their premiums—that mental disease was too clinically ambiguous to insure responsibly.

The fear was hardly irrational. The theory of psychotherapy held that everybody could benefit from it. So how could insurance cover it? In fact, one of the earliest private insurance plans did cover psychotherapy; it promptly went bankrupt. More serious psychiatric disorders—the type that would have to be treated in hospitals—represented even more financial vulnerability for insurers. In diagnosing psychiatric illness, physicians had no lab tests and very few observable clues to use as reference; they had to rely heavily, in some cases exclusively, on descriptions by the patients themselves. Also, treatments were less exact. In other words, since the science of psychiatry was inherently less exact than the science of physical illness, the moral hazard—the potential for patients to seek care they really didn't need—was much higher. And since mental illness had traditionally been a responsibility of government anyway, the private health care sector was more than willing to leave it that way.

In the 1960s, however, private insurers started to add mental health benefits to their comprehensive plans. Although the benefits were frequently inferior, with higher co-payments and stricter limits on overall expenditures per beneficiary, most people with job-based insurance now had them. By 1983, according to the Health Insurance

Association of America, 99 percent of corporate group health policies had acquired at least some mental health coverage.

A significant reason for this was the demands of employees. Thanks in part to the availability of better drugs to treat once disabling conditions such as depression, schizophrenia, and bipolar disorder (also known as manic-depressive illness), people had become far more willing to talk about their mental afflictions—and seek treatment. But businesses that invested in mental health treatment were also acting out of self-interest: workers with mental health problems, particularly those involving addiction, were costing the business community money through lost productivity. Rather than fire and then replace these workers, employers decided that it was better to help them get help. In 1981, an official at Xerox boasted about the returns the company's alcoholism program was yielding: "Recent studies show that some companies are getting a return of up to $6 for every dollar spent on such treatment. Instead of terminating people, companies get back healthy employees. They buy back half the time they were not getting in the first place and save part of the expense of recruiting and training new people."

Then came the 1980s—and the skyrocketing health insurance premiums that sent employers (and, eventually, everybody) into a panic. Scrutiny immediately turned to psychiatry, because the cost of treating psychiatric illness had been rising even faster than the cost of treating physical illness—and not necessarily for the right reasons. On close inspection, the single biggest source of higher costs during the 1970s and 1980s turned out to be a dramatic rise in the admission of juveniles to mental hospitals. And it soon became apparent, first from anecdotes in newspapers and then from statistics compiled by researchers, that parents were increasingly using psychiatric institutions for kids who weren't truly mentally ill, at least not seriously. "Parents used to take problem children to ministers and priests," a researcher at the University of Michigan concluded after studying data on admissions. "Now they take them to psychiatric experts."

Of course, the parents hadn't come up with this idea on their own. The increases in admissions of adolescents and in addiction clinics— the other big source of higher costs—followed after for-profit facilities had targeted such clients aggressively through advertising and marketing: the number of facilities for adolescents and for people with addictions had doubled from 220 to 440 between 1984 and 1988.

It wasn't hard to see why there was such an opportunity to make a profit. In addition to all the businesses now interested in treating their workers' addictions, states had been passing laws requiring insurers to cover juvenile psychiatric care. And if those mandates generally reflected the best of intentions, they often led hospitals to make outreach efforts on less than good faith. In Texas, hospitals were hiring recruiters—and paying them "bounty fees"—for every addiction patient they could enroll in the programs. In Florida, hospitals were investigated for institutionalizing children improperly.

As one of Wall Street's most respected analysts said in 1986, "Psychiatric care is obviously out of control." But it would not remain that way for long.

———

Looking back on the early days of their relationship, Russ Doren now detects many clues that Gina had serious psychiatric problems. But he didn't recognize them at the time—maybe because they were not so easy to spot as they are in hindsight, or maybe because he was just too smitten to notice. It was 1977 when they first met. Russ was twenty-six years old and, by his own account, not much of a ladies' man. Six-foot-four and rail-thin, with angular facial features, he was shy with the opposite sex—the kind of guy who would sit at the phone for hours waiting for the right time to call and ask a girl out. So when his sister offered to set him up with a friend of a friend, he readily agreed,

even though it meant coming to a barbecue at his sister's house in
Cheyenne, Wyoming—two hours away from his home in Denver.

One glimpse of Gina convinced Russ that the drive had been worth
it. She was tall, like him—a full six feet—with green eyes and feath-
ered auburn hair that fell down just short of her shoulders. She looked
like a model, Russ thought, and it turned out that as a teenager, she'd
appeared in some fashion shows. But despite her good looks Russ
found Gina easy to approach; though hardly shy, she wasn't at all con-
ceited and seemed genuinely flattered to have Russ's attention. After
the cookout they went dancing at a local hotel bar; and a few hours
of disco later, Russ felt so comfortable with her that he asked her out
again—even though it would mean another four hours of driving to
and from Cheyenne. She said yes.

Gina, then nineteen years old, was living with her mother and
stepfather while she attended cosmetology classes. But it quickly
became apparent to Russ that Gina was eager to get out. On their
first date, according to Russ, Gina and her mother had gotten into
a loud, rancorous argument in the living room, right in front of
him. (It apparently began when Gina's mother made some comments
about religion to Russ, who is Jewish.) On their second date, also in
Cheyenne, Russ had carefully planned out dinner and a movie there.
But Gina had different ideas. She suggested they drive to Denver.
Russ was pretty sure Gina didn't mean that she would be staying
in Denver overnight. (On the last date, when he awkwardly tried to
make a move on her, she didn't take it too well—warning him, half-
jokingly, that she would punch him in the face if he got any closer,
and then ranting about men who thought they could have their way
on a first date.) So this would mean two round-trips to Cheyenne for
him—eight hours of driving. But Russ was hooked at this point, as
much by Gina's independence and impulsiveness as by her looks. On
to Denver they drove.

It was the first of many such dates the couple would have. Indeed, for several months, Russ and Gina saw each other every weekend; and, almost every time, Russ would end up making two round-trips— to Cheyenne to get Gina, back to Denver with her, back to Cheyenne to drive her home, and then back to Denver. By now, Russ had surmised that Gina's desperation to get out of Cheyenne had something to do with her family. But he didn't ask; and it was Gina herself who finally brought up the subject, explaining one night that she and her mother had been fighting for years—and that, while still a young teen, she had repeatedly run away from home. One time, Gina explained, she was so upset that she intentionally overdosed on a bottle of Tylenol. Russ thought he understood; based on his own limited exposure, he could see that relations between the two were strained. So he didn't ask about it again.

Instead, Russ was focused on a different problem: fatigue. The drive was pleasant enough when Gina was in the car. But he was on his own half the time—including the last leg, which was late at night. To keep from falling asleep, sometimes he would drive barefoot. When that didn't work, he would open the windows and let in the cold mountain air. When he was truly desperate, he'd stick his feet or even his head out the window for a few seconds, just to jolt himself into a more alert state. But even that strategy had its limits, particularly when snow made the drive even longer. One night, after one of the winter's worst storms, Russ didn't return home until 5:30 a.m. And it was a school day. He showed up at work having slept just twenty minutes and sick to his stomach. He decided that this was a good day to show his students a movie. When the loud part came, he went to the sink in the back room and "puked my guts out."

That night, Russ called Gina and decided to take the initiative. "I'm falling asleep at the wheel; I'm almost driving off the road," Russ said, launching into a monologue. Gina apparently thought he wanted to break up, but Russ had something different in mind: "Let's just get married." After a rare minute of stunned silence, Gina piped up, "Great idea." A few months later, the two got married in a mostly nondenominational ceremony presided over by a Baptist minister. (Russ and Gina figured that in this way they would satisfy Gina's parents, without committing themselves to a religious tradition.) Then it was off to a honeymoon at California's Disneyland.

As a newlywed, Russ was as happy as he could ever remember. He loved the way Gina had taught him to be whimsical, like the time she had told him she wanted to go camping—that night. They were living in a mobile home at the time, and the next thing Russ knew, they were tossing their foam bedroom mattress out the window so that they could spend the evening in their garden, lying under the stars. Gina also liked to eat at odd hours, such as 2 a.m. She wasn't pregnant—just hungry. And Russ, as lovestruck as when the two had first met, would oblige, becoming a regular late-night patron at the local Taco Bell drive-in.

Still, Gina and Russ had known each other just a few months when they'd gotten engaged. It wasn't for another two years, apparently, that Gina finally began to trust her husband completely—and started opening up about her past. It happened in the middle of one night, when Russ awoke to a shake of the shoulder, the kind that usually precipitated a taco run. But Russ quickly noticed that Gina's eyes were swollen and moist; she had been crying. "Russ, I have to tell you some things I've never told anybody outside of my family." Then she finally told him the real reason she'd had such problems back in Cheyenne: she'd been sexually abused as a child.

Over the next year, according to Russ, Gina painted a fuller picture of the pattern of abuse—about how her abuser would slip alcohol into her juice, get her woozy, then assault her while her mother was away; and about how she struggled to convince her mother that it was really happening. The abuse was the reason Gina had started running away at such a young age: she had wanted to find her biological father, whom she hadn't seen since he and her mother had divorced, when Gina was seven. But now that Gina really was away from Cheyenne, for good—and now that she finally felt secure in a relationship of her own—she was starting to feel better. In fact, she wanted to start a family. Russ did, too.

In June 1981, when Gina went into labor, he thought it would be the happiest moment of his life. But the labor was difficult. Gina had gained an unusually large amount of weight during the pregnancy, in part because the baby was huge. When Kory was born, at nearly thirteen pounds, doctors whisked him away to the ICU, where he would remain for ten days. He would eventually be just fine, but Gina was devastated. "My God, I can't even have a baby right," she told Russ. Back home, Gina became increasingly downbeat and lethargic. At first, it looked like a case of postpartum depression. But then Gina slipped into what is known as a regression, curling up into a tight ball on the bed and shutting out the world. Clearly something else was going on.

The psychiatrist who had treated Gina when she overdosed on the Tylenol as a teenager was still practicing—and he was practicing in Denver. (Although she lived in Cheyenne, Gina had been hospitalized in Denver during that first overdose.) When Gina agreed to begin seeing him again, Russ was hopeful that she'd be back to normal soon. Within a month, however, the psychiatrist called Russ to discuss the situation. Gina was very, very sick, he said. The abuse she'd endured as a child had wreaked havoc on her developing psyche. Even today, the doctor said, she still blamed herself for letting it happen. She needed more intensive counseling and drugs. And she needed to be under strict observation for a while. Gina would have to go into the hospital, he said.

Gina agreed and was admitted to Bethesda Psychiatric Hospital, where she stayed for the next six months—the first of four admissions she would have there between 1982 and 1984. She progressed in fits and starts, with frequent setbacks, typically linked to some interaction with her parents. Gina's psychiatrist had advised her not to have contact with them, but they would call the house occasionally. Whenever they got Gina on the phone, she was a mess afterward, crying for days and usually ending up in another regression. Sometimes she even tried to hurt herself, by cutting her skin with broken glass or burning herself with cigarettes.

The strain inevitably got to Russ, who worried about what Gina's disease was doing both to Gina and to young Kory. (Eventually Russ would start getting counseling for himself.) About the only way Russ considered himself fortunate was in the financial sense: his insurance was still covering most of the hospital and doctor bills during the first few years. But soon that situation would start to deteriorate, too.

———

It would be hard to overstate how dramatically health insurance for mental illness was shifting during the late 1980s. Spooked by the huge price tag for psychiatric treatment during previous years—and angry over the apparent bilking by for-profit psychiatric hospitals that was the main cause—employers demanded action. And they got it. Insurance companies began offering policies that still covered mental illness, but only with very tight restrictions, frequently limiting beneficiaries to no more than thirty or forty-five days a year in the hospital.

In addition, insurers started applying managed care to mental health coverage, restricting beneficiaries to limited networks of therapists and subjecting decisions about treatment to outside review. These were precisely the kinds of changes that would sweep through insurance for all ailments, physical and mental, in the years to come. But mental health care was much more instantly pliable. It was much easier to be stingy about benefits when, as in the case of mental illness, the disease had a certain second-class status. And it was much easier to second-guess physicians when, as in the case of psychiatry, their clinical decisions seemed more inherently subjective. Research showed that managed care approved hospital stays for psychiatry at a rate of just over 50 percent, compared with 90 percent for obstetrics. Under managed care, average hospital stays in psychiatry fell by 47 percent; for other fields, they fell by just 23 percent.

Once again, a new line of business emerged to take advantage of the new opportunity to make a profit—but this time, instead of for-profit hospitals that provided actual psychiatric care, it was for-profit consulting and insurance firms that specialized in managing psychiatric benefits. And they delivered some impressive results. In 1989, one such firm, AmericanPsych Management, immediately slashed the cost of mental health benefits for its clients Martin Marietta and Georgia Pacific by 20 percent. Such savings would prove typical of the new environment for mental health. IBM, Chevron, Du Pont, and Federal Express were among the companies that would go on to realize savings of between 30 and 50 percent in mental health care in the late 1980s and 1990s.

The first health care providers pinched by the new cost-consciousness were the ones who had been most responsible for the sudden explosion in costs years before: for-profit hospital chains. And few psychiatrists shed many tears over this—because they, too, believed the for-profit mental health industry was frittering away money that might be better spent on people who needed more treatment, as well as giving the whole field of psychiatry a reputation for shoddy care.

But the insurance industry's new approach to mental health benefits hadn't singled out those institutions and systems that had exploited the previous insurance system for profit. On the contrary, the hyperconsciousness about cost applied to all psychiatric care—even those areas in which costs hadn't been rising very quickly.

And one of those areas in particular, the treatment of acute or long-term psychiatric disease, seemed particularly ill-suited for aggressive intervention by managed care into the medical process. Given the need for extreme confidentiality in psychiatry, frequently because patients themselves feared the stigma of mental illness, how were doctors supposed to feel about discussing the intimate details of cases with continually changing panels of the insurance companies' bureaucrats—particularly if employers, who were paying the insurance bills, had access to some of the same information? Another consequence of managed care was the shuttling of psychiatrists in and out of approved provider networks with each new year of contract negotiations. But switching doctors is much more difficult with mental illness than physical illness, because it can take months or years for a psychiatrist to gain a patient's trust—and because it can take that long for a therapist to understand a patient's problems.

Most proponents of managed care proclaimed that the increased scrutiny of decisions about treatment—coupled with the tight limits on inpatient care—would merely make the mental health system more efficient, realizing the traditional ideal of managed care: more for less, as pioneered in Oklahoma and on the west coast during the 1920s and 1930s. And years later, some studies found that overall access to all mental health care had actually increased slightly during the early years of managed care, apparently fulfilling this promise. But that was an overall figure for access—and it was probably linked to either the declining stigma of mental illness or the rapid spread of psychiatric drugs, which insurance was generally willing to cover. (In the eyes of many insurers, psychiatric pharmaceuticals were a cheaper alternative to hospitalization.) It was proof, in other words, that people with minor to modest psychiatric problems had an easier time getting treatment.

By contrast, when it came to the worst illnesses—the ones that required hospital care—psychiatrists complained loudly that they were being forced to skimp on treatment, putting their patients at risk. "It's a disaster," a psychiatrist in Connecticut under contract with CIGNA told a reporter from the *Hartford Courant.* "HMOs are involved in the rape of the mentally ill in a way that hasn't been done since the turn of the century." And while such hysteria was perhaps to be expected from the doctors, even a few officials in the insurance industry acknowledged that some of the new restrictions—particularly the limits on days in the hospital—might jeopardize the health of a small minority of patients, inevitably the ones with the most severe medical problems.

In the early 1990s, the medical director for Aetna's psychiatric management unit admitted that the lifetime limit of $75,000 many companies had imposed was probably insufficient for about 10 percent of their workers—and that although companies themselves frequently relaxed the rules for some of the remainder, a few people with "protracted, chronic, intermittent, severe psychiatric disorders" would simply run out of coverage. "They would not have any recourse and may require referral," he said, meaning that they would have to turn to the government for help. In 1987, a private corporate benefit manager was even blunter about the impact of managed care—and what he saw as his responsibility to ameliorate it: "It will be difficult when a person [suffering] legitimately from something serious such as sexual abuse hits that limit. But my job is to help the company manage costs."

One reason such limits weren't more controversial was the perceived hopelessness of the most severe cases of mental illness. But psychiatry had come a long way since the days of *One Flew Over the Cuckoo's Nest*, and Gina was proof that even the most severe cases were frequently less hopeless than they seemed. After four hospitalizations between 1981 and 1984, she finally turned a corner, apparently because she had begun to accept the fact that she was not responsible for having been abused as a child. During the next few years, up until 1988, she stayed out of the hospital entirely and had newfound energy. She quit smoking and started riding an exercise bike. Although she continued to have trouble keeping a job—a problem her psychiatrist blamed on her difficulty dealing with authority figures—she returned to community college and even joined the honor society, eventually graduating with a two-year certificate. At night, she'd take long walks with Russ, talking about what life would be like when they got old. On weekends, she'd pile onto bean-bag chairs with Russ and Kory for "bad video night." *Attack of the Killer Tomatoes* was a family favorite.

Then, one day in October 1989, Gina's mother telephoned. While she understood that Gina wanted nothing to do with her anymore, she wanted to see her grandson, Kory. For a month, Russ became the go-between for Gina, her mother, and her psychiatrist, until they agreed on a plan. Russ would take Kory to meet Gina's mother and stepfather at a shopping mall, while Gina stayed home. The visit went fine. But when Russ got home, Gina was sobbing, apparently full of guilt for having kept Kory away from his grandmother. Another descent into depression followed, and a week later her psychiatrist readmitted her to the hospital. This was the first of a dozen admissions that would take place over the next year.

Now, though, Russ had an additional worry: the insurance. By this time, the Cherry Creek School District had switched policies. The new policy covered only forty-five days of hospital care each year, at 50 percent. And the Dorens exhausted that limit after the first hospitalization, which ran through the end of February. Before, Gina wouldn't come out of the hospital until she was close to being self-reliant. But now, Russ would later say, there was constant pressure to move her out of the hospital more quickly—perhaps before she was ready. On one afternoon shortly after a discharge, Gina showed up at Russ's classroom, looking very pale. "I've taken all of my medications," she

said. "I love you. This is good-bye." Russ drove her to the hospital
emergency room, which was just a few miles away, where doctors
pulled her through after pumping her stomach. Another afternoon,
Russ came home to find Gina lying on a sleeping bag in the garage,
clutching a bear he had once given her; the key was in the ignition, but
the car, apparently, had stalled after running for a few minutes. Once
again, an ambulance took her to the hospital. Once again, she survived—this time after treatment in a hyperbaric chamber to raise the
level of oxygen in her body.

The mounting debt weighed heavily not just on Russ but also on
Gina: Gina was constantly apologizing for the money problems she
had caused the family; and on a note written after one of her suicide
attempts, she scribbled, "No more debt. No more hospitals." With
the encouragement of Porter's staff, Russ eventually decided to get
help from the government. But the government was in no position
to help. All across the country, the cutting back of private insurance
benefits for mental health had shifted more responsibility for psychiatric care to the government—in effect, reversing the evolution that
had taken place over the previous twenty-five years. But this period
was unlike, say, the Kennedy era; now, neither Washington nor the
states would pump large amounts of new money into mental health.
That left the government insurance programs and government hospitals, which were never adequately funded in the first place, even more
overwhelmed. And Colorado's public mental health system was no
exception.

At the time, Colorado owned and operated two mental hospitals,
both of them vestiges of the era of state institution building. One of
them was exclusively for criminals. The other, Fort Logan, was for
people like Gina—people who needed intensive, perhaps long-term
hospitalization for acute psychiatric problems but had run out of
money to pay for it. The social workers at Porter, whom Russ had

found both helpful and sympathetic, had suggested that he look into transferring Gina to Fort Logan as soon as it became apparent his insurance had run out. And when Russ visited the facility, a former military barracks and parade training ground that had been retooled as a mental hospital in 1961, he had been impressed: it was clean and soothing in appearance, the kind of place where he imagined Gina might feel comfortable.

At first, Gina was wary of the transfer, because her psychiatrist— who had now treated her for a decade—had no privileges there. Moving to Fort Logan, in other words, would mean starting anew with a state psychiatrist, with no way of knowing in advance if she would like the new doctor. Russ figured he'd eventually persuade her that Fort Logan was better than nothing. And he was right: she eventually relented. But by August 1990, when Gina's psychiatrist wrote the letter recommending her transfer, another, more fundamental problem came to light: Fort Logan had no beds available. It was full of patients. Gina would have to go on a waiting list. And there was no telling when she might finally get to the top of it.

The other potential source of government relief was government insurance, in the form of Medicaid. The social workers recommended this, too. And Russ dutifully followed through, driving to the state assistance office in Denver. There he stood in a series of long lines, wondering how he, with his master's degree in biology and his plum teaching job in one of the state's best school districts, had ended up here. But this trip, too, proved unfulfilling. He left the office with yet more incomplete paperwork: he needed legal proof of both Gina's medical condition and the couple's diminished financial status. Then he'd have to mail it all in. And wait some more while the state bureaucracy pondered Gina's case.

Depression and suicidal tendencies are common among the victims of childhood sexual abuse, because they tend to blame themselves for what happened to them. Also common is the condition called dissociative personality disorder. When children are being abused, they sometimes invent alternate identities—usually, more than one—which they can inhabit, in order to insulate themselves emotionally. Even if the identities recede into the subconscious, they can reemerge later in life, seemingly out of nowhere. This is exactly what started happening with Gina. One day, when Russ arrived at the hospital for one of his daily visits, Gina extended her arm for a handshake, very businesslike. "You must be Russ," she said. "It's nice to meet you. I've heard a lot about you." She then introduced herself as a man named "Lloyd." Gina's other personalities were women, including an affectionate one named "Linda" and a hostile one—Russ never got her name, because she was screaming violently at him—who apparently hated all men. Russ had never seen anything like this. "I can't imagine Gina's situation getting any worse," he said at the time.

And yet it was not long after this that the conversation at Porter hospital took place, the one where the staff told Russ he had to take Gina home—where, they hoped, she might enter a day program. But Gina didn't want to enter a day program. Instead, she opted to register for new college classes and to take a job at the local Arby's. The Dorens had purchased a new home during the 1980s, in the suburb of Aurora, and Gina seemed to be holding her own there, at least at first. But then she decided to drive to Cheyenne overnight and try to work through her problems with her family. The meeting did not produce any breakthroughs. But while she was there, Gina had gotten in touch with her former family physician—apparently through her stepfather. She told the physician that she'd left her psychiatric medications back in Denver and got him to write refill prescriptions.

At home, Russ had long since removed all prescriptions (along with sharp objects) as a precaution against future suicide attempts.

But Russ didn't know anything about the pills Gina had gotten in Cheyenne. One afternoon in late September, Russ returned from work to find Gina falling off the couch, barely breathing. Her face was thick with makeup. (Russ later called it "a damn death mask.") When he rolled her completely onto the floor, he found the bottle of pills, half empty. Gina stopped breathing completely while the paramedics were on the way; and once they arrived Russ began swaying back and forth, crying, as they revived her. Eventually the paramedics called in a "life flight" and, with neighbors snapping their cameras while Kory looked on from the front doorway, the helicopter landed on the street. Then, with Gina aboard, it hurriedly took off for a familiar destination, Porter Hospital, only this time Gina would be going to the medical wing instead of the psychiatric unit. Saving her mind would have to come later. Right now, somebody had to save her life.

After some of the past suicide attempts, Russ had started to wonder what Gina's true goal had been. Some people who try to kill themselves really wish to be dead. Others do not—or, at least, they may not be certain. In some cases, they may be trying to call attention to their plight, to hurt themselves, or to hurt somebody else. Russ had eventually become convinced that perhaps Gina fell into one of those latter categories. If she had really wanted to kill herself with the first overdose in the 1980s, why would she have showed up at the school and told Russ about it? And why wouldn't she have started the car in the garage earlier, so that she'd be dead by the time Russ got home? This time, Russ noticed, Gina had taken the same pills she had taken before—and exactly the same amount. And according to what the doctors had told him, she must have taken the drugs late in the afternoon, figuring that he would come home in time to save her. She wasn't really ready to kill herself, Russ told himself. She wanted to hurt herself or hurt her parents or both, but she wanted to live.

That may have been the case. But Gina had lost some weight since the previous attempt with the pills—and now she seemed to be having a much harder time pulling out of the crisis. Russ visited her at the hospital every day, staring at her withered, tube-ridden body while she lay unconscious; he was mesmerized by the sound of the breathing machine expanding and contracting. Sometimes he would take Kory. And sometimes he would go by himself, lifting her hand to his cheek so that his tears touched her skin.

About five days into her hospitalization, as the drug levels in her body finally fell, Gina regained some consciousness. The tubes in her throat prevented her from talking, but she could write and the nurses had given her a pad. "Does Kory know?" she wrote down. Yes, Russ nodded. "I'm never going to see my mother again," she wrote next, turning angry for a moment. Russ nodded again. Then she wrote one more thing, though she was crying by the end of it: "Russ, I'm afraid that I'm going to die." Emotion welled up in Russ's body. She really didn't want to kill herself—or, at least, she didn't anymore. He hugged her. "Honey," he said, "I promise I will get you through."

But that was one promise Russ could not keep. Five days later, when Russ and Kory stepped out of the hospital elevator to visit Gina, the chaplain and the medical director of the ICU were standing there, waiting for them. "Russ, I'm very sorry," the doctor said, putting his arm around Russ's shoulder. "Your wife has died. There is nothing we could do."

It took a few hours before Russ's parents could arrive. When they did, he left Kory with them and went back to the ICU, drawing back the white curtain to look at Gina one last time. Her hands and lips were blue, her face puffy, her extremities cold to the touch. Russ climbed into the bed next to her, trying to warm up her hands with his. But when he opened Gina's eyelids, he remembers, "there was a blankness, there was a nothingness about her eyes," and that's when he finally came to grips with the idea that she really wasn't coming back. Crying some more, he stood back up and promised Gina that he would do his very best to raise Kory. Then he walked out the door, pausing one last time to look back at Gina and tell her, one last time, "I love you."

———

Years later, the Colorado legislature would consider passing a law that would require insurers to provide equal benefits for mental and physical conditions. Most of the states were also considering such laws, as was the federal government. Mental health parity laws, as they became known, had almost universal support among psychiatrists and other professionals who delivered mental health care, as well as the organizations—such as the National Alliance for the Mentally Ill—that spoke out for people who suffered from psychiatric disease. But these laws also inspired strong opposition, primarily from the business community, which feared that mandating more generous mental health benefits would simply recapitulate the problems of the 1980s—setting off a frenzy of unnecessary psychiatric treatment and jacking up insurance premiums to unbearable levels.

Colorado was one of many states in which individuals whose lives had been affected by mental illness spoke out in favor of parity. Russ was among them, giving a short speech to a state legislative committee. And perhaps because of such testimony, Colorado was among the two dozen states to pass a parity law. In 1996, the federal government enacted a version of parity, too—in large part because Senator Pete Domenici of New Mexico and Senator Paul Wellstone of Minnesota had teamed up to get the law through Congress. Domenici, a conservative Republican, and Wellstone, a liberal Democrat, were unlikely allies, particularly on the issue of health care. But in this case a shared experience overcame their ideological divide: each man had a close relative with serious mental illness.

Not long after the laws had passed, it became clear that all the fears industry lobbyists had used to fight the measure—that enacting parity would lead to unnecessary psychiatric care and out-of-control health insurance premiums—were unfounded. Study after study found that parity barely affected health premiums, whether at the state or the national level. But neither was it clear that parity was actually doing much good. The laws typically said nothing about how insurers judged medical necessity, leaving psychiatrists vulnerable to as much second-guessing as before. Most of the laws also had huge loopholes. Under the federal law, for instance, insurers could not impose a lower dollar limit on benefits for mental health than on benefits for physical health. But the insurers could still impose limits on the number of inpatient days—the limits that had caused the Dorens, and so many other families, the most trouble.

The proponents of parity were aware of these loopholes from the very beginning, resigning themselves to such compromises as the only way to push the law through the legislative process. And in Washington, anyway, reformers had hoped that the initial parity law would be a stepping-stone to stronger regulation. President Clinton said that he supported such efforts and so, on several occasions, did his successor, President Bush. But in 2002, hopes for the measure faded when it lost one of its champions, Wellstone, to a plane crash during the final days of his campaign for reelection to the Senate. After that, the opposition of interest groups and conservative ideologues—many of whom saw parity as another unnecessary intrusion by government into the practice of medicine—kept the proposal buried in a congressional committee.

Colorado's law was actually a bit stronger than the national version. And Russ was pleased with his small role in advancing the cause. He took Gina's death hard, going through counseling himself, and eventually dealt with it in part by speaking out. He started giving annual talks to local high school students about suicide, recounting his own story and then mixing in statistics about the prevalence of mental illness plus advice on how to get help. He also lobbied the Cherry Creek School District to improve its mental health benefits.

But in some ways, Cherry Creek proved a tougher sell than the Colorado legislature. When he wrote to the board of education a few months after Gina's death, suggesting that it return mental health benefits to something resembling the level of the early 1980s, an official wrote back, "Unfortunately it isn't possible for us to do everything we would hope to do for our employees. I am aware that the district coverage for mental illness was reduced. This was necessary because of the high cost the school district was assuming for covering all employees for this eventuality. The cost to the district was prohibitive."

It reminded Russ of the reaction he'd gotten before, while Gina was still alive and he repeatedly urged the school board to get better mental health benefits. "I'd go to a committee meeting," Russ later recalled. "They would say, 'I'm sorry, but there aren't that many people who have those kinds of serious problems, and we shouldn't have to burden the rest of us because of the issues of the few.'"

Russ thought he understood what those school board officials were thinking. To them, paying more money for psychiatric coverage meant paying more money to treat an illness they would never have. Mental health, in other words, was somebody else's problem. Russ figures he might have thought the same thing, if only he didn't already know better.

CONCLUSION

WASHINGTON

In many ways, public opinion about health care in America remains as ambivalent as it was at the end of the Clinton health care fight. And maybe even a little more so. If you go around the country and ask Americans whether they like their country's health care system, the majority will frequently say "no," depending on how you ask the question. But the public's true feelings are too complicated for such a simple survey question to capture. Most Americans still have health insurance and, more often than not, they're reasonably happy with it. They know that millions of Americans aren't so lucky, but, according to the polls, they have a hard time imagining themselves in that situation. (They continue to believe, for example, that most people without insurance are unemployed.) Similarly, while they grasp that the uninsured don't always get the same quality care as people with coverage, they still think the uninsured get medical attention when they need it.

But then how do they explain what happened to the Rotzler family of Gilbertsville? Gary supposedly did all the right things to realize the

American dream. He got a college education, acquired the kinds of sophisticated skills the high-tech economy prizes, and went to work for a growing multinational corporation. Even when the pink slips came, he continued working, stringing together part-time work until his old employer hired him back. But because he came back as an independent contractor, not a full-time employee, he was able to replace only his income, not his health insurance. And he paid dearly for that when his wife developed cancer.

Janice Ramsey, the realtor from central Florida, had founded a home construction company with her husband. But the American health care system didn't reward her for entrepreneurial spirit. It punished her. Without access to a large group insurance plan, she had no chance for coverage except to buy it from a carrier that sold directly to individuals and small businesses. Those carriers screened for high medical risks. And Janice, a diabetic, was exactly the kind of person they didn't want to cover. Her futile search for affordable insurance led her, finally, to a scam that left her with a five-figure debt, a damaged credit rating, and no insurance to cover future diabetic complications.

Elizabeth Hilsabeck's husband had a college degree and a successful career at a local bank—successful enough to buy them a home in one of Austin's tonier suburbs. But when Elizabeth wanted to get physical therapy for her severely disabled baby boy, so that he might learn to walk, the couple's HMO said no, even though the therapy was a standard treatment for the disease. Parker Hilsabeck walks today, but that may be only because Elizabeth was willing to sacrifice her house—and, ultimately, her marriage—in order to pay for his care.

In America, a safety net is supposed to exist to help people like this. But in the last few years, people who have relied on the safety net have discovered that it won't always be there for them. Lester Sampson of Sioux Falls is one of them. Even with Medicare, he needed supplementary insurance to cover the full expenses for him and his wife. After putting in thirty years as a meatpacker, Sampson thought he could count on his former employer to provide it. He was wrong.

Sampson made out all right in the end, all things considered. But not everybody has been so lucky. Too poor to pay for his drugs on his own, Ernie Maldonado was counting on Tennessee's Medicaid program to cover the out-of-pocket costs associated with treating his many health problems. When Tennessee decided that it could no longer afford to be so generous, Ernie stopped taking his drugs. A few weeks later, he died.

As a last resort, people without a way to pay for their medical care can still turn to charity and public hospitals, which care for the poor and uninsured just as they did 100 years ago. But those institutions are not able—and, in some cases, not willing—to shoulder the burden. The Los Angeles health network may be extremely dysfunctional, but nobody thinks that any big-city public system has the resources to care adequately for millions of uninsured people. If Los Angeles had such resources, Tony Montenegro might have kept his diabetes under control and maintained his ability to provide for his family.

Nor are private providers of charity care an adequate substitute. Many treat the uninsured grudgingly. They also expect to be paid—a lot. Indeed, as Marijon Binder discovered in Chicago, not even a former Catholic nun receiving care at a Catholic hospital can count on much compassion at a time when nonprofit hospitals feel they must demonstrate a for-profit mentality.

And then there was Russ Doren, who over the course of a decade confronted virtually all of these problems. He had not just a college degree but a master's, too, plus a teaching job in an affluent suburban school district. But that wasn't enough to protect his family when his wife, Gina, developed severe psychiatric problems. First private insurance failed them. Then the safety net did. Now Gina is gone.

Not everybody who is uninsured or underinsured ends up suffering serious medical or financial consequences. At any one time, only a small percentage of people will have severe health problems. In this sense, the stories in this book are not so much representative as indicative—indicative of what can happen even to hardworking, intelligent people when their need for medical care overwhelms their ability to pay for it. But the whole point of insurance is to protect against a misfortune that, however unlikely, would be catastrophic if it struck. As these stories show, the risk is greater than most Americans seem to realize.

And the risk is growing. Ever since modern health insurance appeared in the early 1920s, large numbers of Americans have gone without its protection—and suffered serious consequences as a result. But for most of that time, protection was at least expanding, first through the spread of private coverage and later through the creation of Medicare and Medicaid. By the early 1980s, nearly 90 percent of Americans had some form of health insurance. And it was generous insurance, by and large, covering a wide variety of services while imposing relatively small out-of-pocket payments. Meanwhile, even those without coverage could turn to a medical safety net that, if not always eager to accommodate them, nevertheless had resources to provide for large numbers of them.

Now, the evolution is running in reverse. The number of employees getting insurance from their jobs is declining. Individuals trying to buy insurance on their own, always at a disadvantage in a market designed around large groups, will only struggle more if genetic testing and improved information technology allows insurers to become more selective. Public programs can't keep up with the demand for services. At any one time, 16 percent of the population, about 46 million Americans, have no health insurance at all. This is the highest proportion since the 1960s. (According to one projection, by 2013 the figure will be up to 56 million.) And even some of those with secure coverage struggle with their medical bills, because their premiums and out-of-pocket payments are rising faster than their incomes—or because they need services that their insurance doesn't cover. They will become dependent on institutions that provide charity care, where they might get excellent care—or they might wait three days in the emergency room.

It is, in short, a crisis. And it's not one the private sector will, or even can, solve on its own. Companies are increasingly reluctant to pay for their employees' insurance because it's no longer a cost-effective way to keep a stable workforce. Insurers are wary of covering individuals with serious medical problems—and HMOs apply extra scrutiny to their treatments—because those people run up big bills, which the insurers will end up paying. Hospitals try to keep out the uninsured because treating those patients will just divert hospital resources away from paying customers. And while these actions may sometimes seem unsavory, they are actually perfectly rational, entirely predictable reactions to the incentives of the marketplace.

The last time the private sector's failure in health care was this clear-cut was the 1960s, when it became apparent private insurance was not up to the job of covering America's elderly—and the government responded by creating Medicare. But the conservatives who have run the country for most of the last few years haven't sought to bolster existing programs, introduce new ones, or otherwise intrude upon the prerogatives of the health care industry. On the contrary, they have clung to their faith in the private sector. And they have acted accordingly. That is why, as the governor of Texas, President Bush resisted expanding the State Children's Health Insurance Plan, even though federal money was there for the taking and Texas, with relatively more uninsured residents than any other state, desperately needed the help. That is why he refused to sign measures regulating the practices of the state's HMOs, even though the legislature passed them twice. And that is why, in Washington, he has proposed to gut Medicaid just as Ronald Reagan once did. Even the Medicare drug benefit—which many of Bush's fellow Republicans consider a financial monstrosity and unconscionable expansion of the welfare state— is in keeping with this pattern. By turning more of the program over to insurance middlemen, squandering its funding on unnecessary subsidies to health care industries, and enticing its healthier beneficiaries to leave, Bush and his allies have in some respects weakened it.

One reason for this approach is that, at some fundamental level, Bush and his allies don't seem to think the affordability and availability of health care are a major problem that commands their sustained attention. As a result, to the extent they even think about the issue, they see it largely in the same terms that their constituent interest groups do: regulations are bad because they hamper the ability of the insurance industry to operate as it sees fit. A government-run drug benefit for seniors is bad because it would eat into the profits of the pharmaceutical industry. Public insurance programs are bad because they have to be financed with taxes, inevitably imposing the largest penalties on either the wealthy or big business.

That deference to well-heeled constituent groups also helps explain what may turn out to be the most pivotal health care policy change of the Bush era: The creation of Health Savings Accounts (HSAs). HSAs are available to people who forgo traditional health benefits and opt, instead, to buy high-deductible insurance that covers only catastrophic expenses. These people can deposit money into the accounts free of taxes and then withdraw cash, again without taxes, to cover whatever medical expenses the catastrophic policies don't. What they don't spend in a given year they can roll over to the next; upon retirement, they can withdraw it (or pass it along to their heirs).

Insofar as HSAs make less comprehensive insurance seem more

appealing, this represents yet another nod toward Republican con stituent groups—such as employers desperate to shed the burden of financing generous employee benefits. But advocates of HSAs make broader philosophical claims on behalf of their idea. One is that HSAs will foster better, cheaper medical care by turning patients into aggressive consumers who can shop among doctors and hospitals; another is that HSAs will encourage people to live healthier lives, in order to avoid costly future medical bills.

Behind both statements lurks one common, though not always articulated, belief: that the fundamental problem with American health care today is that people have too much insurance. As this argument goes, coverage for relatively minor medical expenses, a standard feature of traditional insurance, encourages people to consume more care than they need. By making people pay more of their expenses directly out of their own pockets—by giving people "more skin in the game" as the saying goes—HSAs combined with catastrophic insurance would reverse that. In other words, it would make people be more responsible.

The insight about the nature of insurance has merit. Many people really are careless, if not downright irresponsible, when it comes to their behavior and consumption of medical services. Nor is this insight entirely new. The idea that insurance insulates people from the true cost of medicine, thereby encouraging waste, was a building block of the managed care movement. What's different about HSAs is where it seeks to extract cost savings. Rather than change the behavior of both the people providing care (doctors and hospitals) and those receiving it (the patients), it focuses its efforts almost exclusively on the latter.

That seems dangerously naïve. How are consumers supposed to shop for good medical care when most experts still don't agree on how to measure it? If poor health or even death isn't enough to discourage somebody from poor diet and exercise habits, why would possibly bigger health care bills at some indefinite point in the future encourage that person to act differently? And who's to say that people won't skimp on the services that might really benefit them, like routine checkups or maintenance care for chronic conditions? Various studies have suggested that's not just possible but likely.

Still, the most pernicious effect of HSAs would be on the broader balance of how we finance health care in America. Traditional, comprehensive insurance has the greatest impact on people with serious medical problems, by subsidizing and thus reducing their direct bills with the premiums that relatively healthy people contribute into large risk pools. The more insurance recedes, the more people with serious medical conditions lose that protection. In other words, the overall burden for financing health care in our country will shift from the healthy to the sick. And particularly as medicine becomes more expensive relative to living standards—something that technology and an aging society quite possibly make inevitable—even middle-class people could find this burden more than they can bear.

This would represent, in a broad sense, a continuation of the evolution that took place throughout the second half of the 20th century—in private insurance, as community rating gave way to experience rating, and even in public programs, as Medicare failed to keep up with rising out-of-pocket expenses and (more recently) as Medicaid failed to keep up with rising numbers of uninsured. Eventually, the situation could even start to resemble the one that prevailed during the late 1920s, before modern insurance came into existence—a time when experts' best advice to people with illness was to be thrifty and save, a time when collective responsibility for the burdens of the sick basically didn't exist, and a time when large portions of the American population were left to face the threats of illness on their own.

But if the future really is starting to resemble the past—if American health care really is reverting to what it looked like before modern insurance existed—then perhaps it's time to reconsider the option the United States rejected back then, and repeatedly since: universal health care. Universal health insurance can take many different forms. On one end of the philosophical spectrum are the systems in which government provides health insurance directly to citizens. These single-payer systems, as the experts call them, exist in Britain, Canada, and Sweden. On the other end of the philosophical spectrum are systems in which private entities continue to deliver insurance, but do so under close government regulation. These hybrid systems come in many varieties. One example is the German health care system, in which people get their insurance through a group of private, nonprofit "sickness funds" linked to their jobs. Another is the system Clinton tried to create here, in which Americans would have purchased coverage from private insurance carriers either through their jobs or through government-run purchasing cooperatives.

Despite all their differences, the significance of these schemes lies in what they have in common. In every universal health care system, the government begins by defining a set of benefits to which everybody is entitled, and then finds a way to make sure everybody gets these benefits. Universal health care isn't free: people must pay for their health insurance through some combination of taxes and premiums, and they frequently owe co-payments on some if not all services. But the government can (and usually does) adjust these required payments depending on people's ability to pay. It can also use leverage to control costs more effectively than a fragmented, non-universal system can. In the hybrid systems, people have many choices regarding type and scope of coverage. (Under Clinton's plan, for example, everybody could have chosen between an HMO and a traditional fee-for-service plan.) But government regulation of marketing, along with the defined benefit packages, theoretically reduces the ability of private insurers to cherry-pick the healthiest enrollees—creating a giant risk pool, much as a single-payer system would, so that the financial burden for medical expenses ultimately falls on the entire population and not too heavily on those individuals with the most health problems.

One way to look at this is to realize that all universal health care systems, even those that rely heavily on private insurance, require the government to "run" medical care. But however unfashionable politically, the idea has at least one powerful piece of evidence going for it: Medicare. Despite the government's reputation for bureaucratic misery, from a beneficiary's standpoint the Medicare program is almost stupidly simple. It allows patients to see any doctor who chooses to accept the fees Medicare sets—in reality, this means the vast majority of doctors. And it allows its beneficiaries—unlike most Americans now in managed care—to receive medical services without going through referrals or preauthorization. In fact, the elderly have far more positive feelings about Medicare than working-age people have about their private insurance. (The new Medicare drug benefit may turn out to be another story, but, of course, that's run by private insurers, not by the government.)

To some experts, the fact that Medicare has pumped so much money into health care—and, as a result, put such a huge drain on the federal treasury—makes it a particularly ill-advised model for delivering health insurance to the rest of America. And, again, there's some truth here. Or, at least, there once was. During its early years, hospitals seized on the new program to finance vast new expansions which, in the days before managed care and insurers' scrutiny of billing, brought in patients for services they really didn't need. But Medicare had inadequate cost controls in part because during the 1960s, its political sponsors worried that doctors and hospitals would kill the program if they felt too threatened by it. Once the program was firmly established, the government introduced more serious cost control; since then, it has actually held down costs better than private insurance.

Medicare, in fact, is remarkably efficient. It doesn't have to invest in marketing or clever actuarial schemes to avoid financially risky beneficiaries. It has no executives on which to lavish seven- or eight-figure salaries and it has no shareholders to whom it is expected to deliver dividends. Medicare has its problems for sure. If anything, the program could probably stand to spend a little more money on administration, to encourage better quality care. But precisely because its overhead is so ridiculously low compared with that of private insurance, it could afford to do so and still provide more insurance bang for the buck.

Even more compelling evidence to support universal health care comes from abroad. No other country in the world comes even close to spending 16 percent of its wealth on medical care, as we do. Yet there is precious little evidence that our extra spending makes us healthier. On relatively crude measures, such as infant mortality and life expectancy, the United States is about average for a highly industrialized nation. On more sophisticated measures such as "potential years of life lost" or "quality adjusted life years," which experts have devised in order to measure specifically the way the health care system works, the United States is actually a little worse than average.

Of course, if you try to engage a critic of universal health care, you probably won't hear about these statistics. You'll hear instead about rationing and waiting lines, followed by a horror story from Britain or Canada. These arguments against universal health care are more potent than most others, because they are at the heart of the public's ambivalence: the fear among more affluent Americans that with universal health care, they will lose the access to medical care they already have.

But the facts simply don't support this. The stories about Canada are wildly exaggerated. And the pinched access to services in Britain, at least, isn't a product of universal health care. It's a product of universal health care on the cheap. The British spend just 7 percent of their national wealth on health care, less than half of what Americans spend. It's possible to spend more than that—and get more—while still spending less than the United States does. A perfect example is Japan. Relative to the United States, Japan spends about 60 percent as much of its wealth on health care. But the Japanese don't wait for medical services. And they have more "stuff." In fact, Japan leads the world in the availability of technology such as CT scanners and MRI machines.

Still, the best showcase for what universal health care can achieve may be France. France provides insurance to its residents through quasi-independent sickness funds, which are overseen by the government. And the insurance covers most health services, including visits to doctors, hospitalization, and prescription drugs. The French finance this insurance through taxes and general government revenue, rather than premiums, and impose substantial cost sharing. But in order to prevent cost sharing from penalizing people with serious medical problems—the way Health Savings Accounts threaten to do—the government limits every individual's out-of-pocket expenses. In addition, the government has identified thirty chronic conditions, such as diabetes and hypertension, for which there is usually no cost sharing, in order to make sure people don't skimp on preventive care that might head off future complications. Private insurance has a place, too: The French can purchase supplementary coverage to pay for what the national health insurance plans don't. (People too poor to buy it on their own can get such supplemental coverage directly from the government.)

Like every country with universal health care, France exerts more

control over the diffusion of medical technology than the United
States does, scrutinizing new treatments and procedures in order to
test whether they add clinical value and are reasonably cost-effective.
This is precisely the kind of control that spooks many Americans,
who fear that creating a universal health insurance program here will
mean destroying access to cutting-edge medical technology. But at a
time when the cost of just one medical episode dwarfs what many
average citizens earn in a year, all health care systems—even ours!—
must make choices between how much they are willing to spend for
medical care and what they expect to get in return. The difference is
that in France, those decisions are made openly and explicitly, by officials who are ultimately accountable to voters, rather than secretly
and implicitly by companies accountable to shareholders. They are
still difficult decisions, to be sure. But in universal health care systems,
they tend to be fairer, more equitable, and simply less painful.

The proof is in the results. The French have easy access to medical
care—easier, in fact, than their American counterparts. A Frenchman
can see whatever doctor he wants whenever he wants, a privilege even
most affluent Americans surrendered long ago, thanks to managed
care. Waiting lists and lines, the supposed drawbacks of universal
health care, appear to be nonexistent. And how good is the medical
care itself? Although they have slightly less technology than we do,
statistically the French seem to do slightly better than Americans on
most measures of health outcomes. (This is yet another reminder that
in health care, more is not always better.) One recent study comparing care in Manhattan and Paris found the care in Paris to be better, largely because the Parisians have better access to routine and
effective primary care—which is typical in universal health insurance
systems. It also turns out that in the international survey of public approval, the one in which the United States finished fourteenth out of
seventeen, the French finished first.

The French health care system isn't perfect. No system is. If you talk to the French, they'll cite some of the same concerns we have here—particularly the uneven quality of medicine and the fact that the demand for technologically advanced care always seems to outpace the desire to pay for it. Yet the French universal health care system seems no less capable of addressing these problems; indeed, the centralization of authority and bargaining power suggests, if anything, that it ought to have an advantage. And while it's possible that more intensive research—the kind that's already been done in Britain, Canada, and Germany (where English is more widely spoken)—will reveal hitherto unreported flaws in France's scheme, it's hard to imagine they would seriously outweigh the system's virtues. That's particularly true when you consider its most unambiguous advantage over the American system: financial barriers to care are virtually nonexistent.

Indeed, the critics who carp about rationing abroad, whether real or imagined, never acknowledge the fact that rationing is a reality in our system, too. The difference is that we ration by income and by medical condition. And that rationing seems likely to get worse with time.

———

Ever since the defeat of Clinton's plan for health care, two groups have dominated the debate over health care in Washington: those who think universal health care would be bad policy and those who think it would be bad politics. And those in the latter group, at least, have good reason to think the way they do. The Clinton fight was every bit the political disaster its critics now claim it was. By 1994, the public had soured not just on the plan but on the whole idea of universal health care. The pounding by special interests opposed to reform had a lot to do with this, of course, but so did the political mood. And in the ten years that followed, Americans handed the Congress and eventually the presidency to Republicans who made very explicit their belief that—as President Bush put it in his 2003 State of the Union address—the system's "problems will not be solved with a nationalized health care system that dictates coverage and rations care."

Still, times change and public moods shift. The mid-to-late 1990s were a period not only of skepticism about the public sector but also of supreme faith in the power of individualism. Anybody could get rich overnight, it seemed, just by logging on to a computer. You didn't even have to start your own company; it was enough just to invest in the right high-tech start-up. This was hardly the first time in American history that such an idea had broad-based currency. (A fellow named Horatio Alger did pretty well writing novels about it in the nineteenth century.) But the seeming power of the Internet to break established rules of the economy helped convince many experts—and, indeed, much of the public—that government-run social insurance programs were obsolete relics of the Great Depression, as hopelessly out of place in the modern economy as Western Union telegrams or the Model T.

Then the Internet bubble burst and the economy fell into recession, as it always does fall eventually. People lost their jobs, and even when the economy began to recover, living standards stagnated. In the 1990s, the culture idealized entrepreneurs and glorified wealth; politics, in turn, dismissed populism with the epithet "class warfare." When Al Gore campaigned for president on a platform of "the people versus the powerful," the opinion class could barely contain its titters. And yet almost immediately after that election the newspapers were filled with stories of corporate corruption, such as the accounting scams at Enron and the infamous profiteering by power companies that left large swaths of California literally in the dark one summer.

Twice before in American history, similar periods of corporate debauchery and free-market hysteria paved the way for a new politics—one founded on the proposition that, by acting together through the government, people could balance out the power of the private

sector and serve their greater, common interests. The first time re-
sulted in the Progressive Era; the second time in the New Deal. Given
the spectacles of corporate excess in the news today—and the deepen-
ing concern over economic insecurity—it is not so far-fetched to suggest that another such period may be on the horizon, particularly as
regards to health care. Politicians now get more mileage from bashing
an HMO than they do from ranting about welfare queens as Ronald
Reagan often did. Middle-class families with kids may not like seeing
the government take taxes out of their paychecks, but if you sit at
their dinner tables, you're more likely to hear them complain about
a pharmaceutical company charging their grandparents thousands of
dollars for life-preserving medication. In fact, criticism of the drug
industry—and big business more generally—was a key factor in the
Democratic Party's landslide win of 2006.

To be sure, it takes more than bogeymen to make large-scale political reform possible. Resentment can stir passions, but it takes hope
to sustain the kind of movement necessary to enact sweeping policy
initiatives. Ronald Reagan's great rhetorical achievement during the
1980s was his ability to marry skepticism about the government to
public aspirations for a greater tomorrow. It may have been twilight
for the New Deal and the Great Society, but, he promised, it was still
"morning in America."

But at least when it comes to health care, the principles of modern conservatism are conspicuously short on such comfort or hope.
Bush and his allies promise that their plans will foster choice and
autonomy, but the essence of their plans is to diminish what security
that insurance still provides—and to shift the burden for high medical
bills onto those with the worst medical problems. To the extent that
conservatives even acknowledge the hardship these initiatives could
impose, they respond with a shrug. That's just the way life works out
sometimes, they seem to suggest, and there isn't a damn thing we can
do about it.

In contrast to such stark indifference, the vision of universal health care—one traditionally articulated by liberals—offers optimism. To believe in universal health care is to believe that we can do more and do better, all at once—that it is possible to have hospitals full of high technology and emergency departments with room for all comers; that it is possible for people to choose their doctors and have a say in their treatments; that it is possible to make the economy more free and more efficient; and that it is possible to do all this for everybody, not just an economically or medically privileged few, in a way we can all find affordable.

One way to look at the stories in this book is to catalog the wrongs done. But a more uplifting, if still bittersweet, perspective would contemplate what might have been. In a universal health care system, Elizabeth Hilsabeck might still be married and living in her dream house. Lester Sampson might still be fishing off his back porch on Lake Madison. Tony Montenegro might still be reading mystery novels. Russ Doren might still be taking long walks with his wife. And the Rotzler children might still have their mother around.

To its critics on the political right, universal health care is an imposition on liberty that weakens individual initiative. But this is the classic bait-and-switch of modern conservatism—to make us forget that in a democracy, the government is merely an expression of our will and resources as a community. Universal health care is really about finding collective strength in our individual vulnerabilities— about helping a family member, a neighbor, or a fellow citizen because, next time, any one of us could be the person who needs help. It isn't about *them*. It is about *us*. One day enough people will realize this to make universal health care a reality. The only question is how many more must learn it firsthand—and suffer the consequences— before that happens.

AFTERWORD

Graeme Frost was a bespectacled twelve-year-old boy with a mop of wispy blond hair and a seemingly innocuous story to tell about health care in America. Three years before, Graeme and his sister had been in a serious automobile accident near their family's Baltimore home. They sustained near-fatal injuries—head trauma left Graeme in a coma for several days— and each spent more than a month in the hospital. The charge for this medical care, including the follow-up rehabilitation, was in the hundreds of thousands of dollars. If not for the Frosts' health insurance, they likely would have been stuck with the entire bill, forcing them into severe financial distress if not outright bankruptcy.

But getting health insurance had been no easy thing for the Frosts. Graeme's father, Halsey, was a woodworker who owned and ran his own business; his mother, Bonnie, had a part-time job at a locally based medical publishing company. The couple said they made between $40,000 and $50,000 a year, which put them right around the median for American families. But, as they would later explain, the family didn't have access to a group plan through a large employer because Halsey was self-employed and Bonnie's work was part-time. The only way to buy insurance was to purchase it directly from an insurer or through a broker, where the prices would be higher—so high, in fact, that they couldn't afford it.

But the Frosts did have access to a different kind of insurance: the kind that came from the government. It was through Maryland's version of the State Children's Health Insurance Plan (S-CHIP), the program financed largely by the government, administered by the states, and established in the late 1990s as a way to help uninsured children whose families were too wealthy to qualify for Medicaid. Over the years, many states had decided to open their S-CHIP programs to some middle-class families for the very simple reason that a lot of these families were having trouble finding health insurance. This was particularly true in northeastern states like Maryland, where the cost of living was higher than in other parts of the country. Families with S-CHIP in Maryland didn't necessarily get "free" health care; more affluent families had to pay premiums, plus out-of-pocket expenses for some of their medical services. But it was still far less costly than private coverage, which is why thousands of families had enrolled. The Frosts were one of those.

S-CHIP was popular in Washington, too. And when the law authorizing the program expired in 2007, lawmakers from both parties got together, agreeing on a new law that would not just renew the program but also expand it, to help even more families take advantage of it. But President Bush and some of his stauncher allies had different ideas. They thought the program had already gotten too big, straying far from its mission of helping poor children. They also worried about what policy experts called the "crowd-out" effect: because at least a few of the families who ended up buying into S-CHIP might otherwise have found a way to obtain private coverage, the program's expansion moved the whole country closer to what Bush and his supporters said was "socialized medicine." Bush said he would renew the program, but only at funding levels that wouldn't keep pace with rising medical expenses. He also proposed to limit the program, by law, to only those families at less than twice the poverty line. Citing concerns that Bush's plans would actually force states to cut people from the program, Congress passed an expansion anyway and dared the president to veto it—which he promptly did.

And that's when Graeme Frost came into the picture. As part of an effort to round up votes for overriding Bush's veto, Democratic leaders in Congress asked Graeme—whom they'd found through a health care advocacy group—to give their weekly radio address. And although political parties use average citizens to make their arguments all the time in Washington, this particular decision provoked a bitter, angry reaction. It began on the Internet, where bloggers began posting information about the Frosts that, supposedly, made the family seem a lot less deserving of help. One

writer discovered that the Frosts owned their own home, and that a house
on the same block had recently sold for several hundred thousand dollars; another discovered that Graeme and his sister attended a high-priced
private school. A conservative columnist drove by the Frost home and discovered what looked like a newish-looking SUV there—apparently, one of
three vehicles the family owned. The coproprietor of a website about health
insurance said he had priced coverage in the area; a family fitting the Frosts'
demographic profile could actually have found insurance, he said, for as
little as $450 a month. Eventually Rush Limbaugh picked up the story, as
did more staid print outlets, like *The Weekly Standard*. "If this is the face
of the 'needy' in America," one writer at the *National Review* concluded,
"then no one is not needy."

Curiously, few if any of the people writing about the Frosts actually
bothered to contact them directly. But reporters for more traditional media
outlets, including *The Baltimore Sun* and *The New York Times* did. And,
combined with information they gleaned from other sources, they painted a
somewhat different picture of the family's situation. It turned out that, even
with the rent from the commercial property Halsey owned as part of his
business, the family's income really did seem to be as they reported—somewhere around $45,000. They'd had to prove it in order to qualify for the
Maryland S-CHIP program. As for the house, they'd bought it two decades
earlier for a measly $55,000. At the time, it was in a blighted, drug-infested
neighborhood. Gentrification had hiked the neighborhood's real estate values, but even now it was probably worth only $250,000 or so. (A rumor
that the family had just installed new granite countertops turned out to
be just plain false.) Yes, the Frost children went to an expensive private
school—on scholarship. And, yes, the family had three vehicles. But one
was a beat-up Ford pickup truck that Halsey used for his business and only
for short hauls. That shiny SUV was a gift from some families at the school.
It seems that they'd heard the children were terrified to ride in cars because
of the accident.

As for the availability of private insurance, Bonnie Frost explained that
she had tried to price private insurance recently; not surprisingly, insurers
wouldn't take them as customers since both Graeme and his sister now had
"preexisting conditions" making them bad medical risks. And while that
might not have been the case before the accident, it turns out that the cheap
policy the insurance blogger quoted wasn't necessarily such a great deal,

either. The blogger, who happened to be an insurance agent himself, never specified what services the policy covered. Given the spotty benefits common among insurers who sell in the non-group market, it's quite possible that if the Frosts had such coverage, the bills would still have overwhelmed them financially, leaving them only somewhat better off than if they'd never had coverage at all.

It was true that the Frosts weren't destitute. But that was precisely why their situation was so striking. Here was a middle-class couple, with decent jobs and a home in their name. But they couldn't get affordable health insurance—at least, not on their own. And, in this respect, they were all too typical of the reality America faced, as the latest Census figures showed that the proportion of Americans without health insurance had hit a new peak—and surveys showed millions more struggling with their insurance bills. Maybe everybody was not needy, as that critic had quipped, but an awful lot of people were—enough to move health care back to the top of the political agenda, right in time for the 2008 presidential campaign.

It was the specter of that election—and the debate over major issues it had already sparked—that best explains why Graeme Frost generated so much controversy. For the first time since the demise of the Clinton health care plan in 1994, mainstream politicians were talking openly about making sure every single American had health insurance. The talk had started in the late fall of 2006, right after the Democrats took control of Congress. That is when Senator Ron Wyden, a liberal Democrat from Oregon, introduced a detailed, serious proposal for universal coverage he called the Healthy Americans Act. Within the next few months, the three leading candidates for the Democratic presidential nomination—Hillary Clinton, John Edwards, and Barack Obama—would introduce their own plans. And it wasn't just Democrats embracing universal coverage: In Massachusetts, outgoing Republican governor Mitt Romney signed a law designed to make sure every resident had coverage; in California, Governor Arnold Schwarzenegger, also a Republican, proposed to make his state next.

No less significant was the outside support these plans were attracting. When Wyden introduced his proposal, he was flanked on one side by Andy Stern, president of the Service Employees International Union. This wasn't so surprising; SEIU is probably the most liberal union in the labor movement, and Stern himself is a longtime advocate for universal coverage. But standing to Wyden's other side was Steve Burd, chief executive officer of Safeway. Outside of the grocery business, Burd had only made national headlines once before—when he'd fought with his labor unions two years earlier, demanding they accept skimpier health insurance benefits. It turns out that experience had a profound effect on Burd. He'd taken on the unions because Wal-Mart was coming to California, and he didn't think he could compete with them since they offered their workers less generous health benefits. So once Burd won his fight with the union, he turned around and began agitating for a universal coverage program—because, he reasoned, businesses like his would be better off if they simply ceded the issue to the government, as employers did in so many other developed countries.

Burd's endorsement turned out to be a harbinger. Soon even well-known opponents of universal coverage, like the National Federation of Independent Businesses, announced they were in favor of some sort of universal coverage. Even Wal-Mart fell in line. In one of the most unlikely pairings of recent political memory, CEO Lee Scott shared a podium with Stern, whose union had made fighting Wal-Mart one of its top priorities in recent years. And while it was hard to know how seriously to take these endorsements, since many of these converts to universal coverage were suspiciously vague about the details of what, exactly, they were endorsing, the announcements were indicative of an unmistakable shift in the opinion of both elite leaders and the public at large: Universal health care was a respectable topic in mainstream politics again.

But it was one thing to talk about giving everybody health insurance, quite another to do so. Plenty of conservatives and business groups had said they supported universal coverage in 1993 and 1994, too, only to abandon the cause once a detailed plan emerged. (Among other things, they decided that various provisions, like the requirement that all businesses contribute towards health insurance costs, didn't sound so good after all.) And, as if to underscore just how pinched the political discussion remained, virtually everybody talking about universal coverage this time around—from the politicians to the CEOs—had in mind the very same model, one decidedly less radical than what many reformers considered ideal.

This new model for achieving universal coverage was known as the "individual mandate" mechanism. The essence of this scheme was a bargain between the government and its citizens. The government would require everybody to obtain insurance, which in most cases would mean buying private coverage; in return, the government promised to make insurance available and affordable to everybody, by prohibiting insurers from excluding or avoiding sick people and by giving subsidies to people for whom insurance was simply too expensive. In some respects, it wasn't so different from what Bill Clinton had proposed in 1994: His plan, too, envisioned most people getting coverage through the private sector. But under the Clinton plan, virtually everybody—even people who were already insured, whether through an employer or some other source—had to start buying their coverage directly, through special cooperatives that the government would set up. In the new schemes, people who already had insurance could keep it if that's what they preferred.

A key rationale for this shift was political: Most people with insurance like their coverage, so if you require them to change, they're more apt to oppose reform—no matter whether the new insurance they'd get is better or more secure than the insurance they already have. But this also meant that the new plans were considerably less ambitious than the '94 plan. One reason the old Clinton plan forced everybody to buy coverage on their own, through the cooperatives, was that it allowed the government to get more heavily involved in regulating the price of insurance and the kind of benefits offered. This leverage would, in theory, give government more ability to reengineer the whole medical care system—and, ultimately, hold down its costs. In the new proposals under discussion in Washington and on the campaign trail, government wouldn't have nearly as much power. That made the plans less treacherous politically, but, arguably, it made them less capable of fixing some of the health care system's underlying problems. Simply put, they would scare people less because they would do less.

Another concession to politics—one the new plans shared with their 1994 predecessor—was the heavy reliance on private health insurance. A wealth of evidence still suggested that a public plan, run by the government, would be more efficient than a system of competing private plans, since even with the most effective regulations in place insurance companies seemed likely to spend a lot of their time competing with each other to avoid the worst medical risks—much as they do now. But selling a public plan meant selling the public on the idea that government works. And while the country's political mood seemed to be drifting further and further away from the hyper-individualistic, anti-government feeling that dominated the 1980s and 1990s, it remained profoundly skeptical of the public sector—as, indeed, Americans have been for most of their history. So even though the most popular insurance plan in America was a government-run plan (Medicare) and even though the most popular system abroad was a government-run plan (France), discussion of creating such a "single-player" system to cover everybody was limited to the undersized liberal bloc in Congress, one fringe presidential candidate, and a handful of left-wing interest groups—which is to say, it wasn't getting much attention at all.

Still, if the new plans lacked the promise of either the old Clinton plan or a bolder government-run scheme, they remained promising. Very promising, in fact. All of the serious proposals made in 2007 envisioned setting a generous level of minimum benefits; all of them would have required insurers to sell to all comers, at one price, regardless of preexisting medical condition. All of the serious proposals also called for creating independent institutes, staffed by medical experts, to make sure new treatments were effective. This held out the possibility of saving at least some money, albeit less than a more ambitious plan might; it also had the potential to promote better health, by reducing unnecessary and frequently dangerous over-treatment that study after study had shown to be widespread in the United States. Financially speaking, the proposals seemed reasonably sound. Their projected costs were in the range of $50 to $120 billion a year, which sounds like a lot of money but, in budgetary terms, is pretty easy to raise. (Typically, the plans claimed some savings from the Bush tax cuts for wealthy people, which were due to expire in a few years, and got the rest through savings and fixed contributions from employers.) And while none of the proposals would move everybody into a Medicare-like system, most called for creating a new government-run plan into which people could enroll voluntarily—a plan that could, someday, become the basis of a true "single-player" system.

Most important of all, though, the new health care proposals promised to cover everybody or at least nearly everybody—offering, in effect, a government-backed guarantee that affordable insurance would always be available, so that nobody would ever have to worry about losing coverage again. If successful, that alone would represent a massive achievement in public policy—the biggest, surely, since the 1960s.

By the summer of 2007, universal health care had even penetrated into popular culture, when *Sicko*, a documentary by the controversial director Michael Moore, hit theaters. Although not particularly nuanced in its discussion of policy, it ably showcased the flaws of American health care, particularly the extent to which even people with insurance—the supposedly lucky ones—found themselves exposed to huge financial and medical risks.

But support for universal coverage was hardly unanimous. Not long after Moore's movie hit the big screen, Republican presidential candidate Rudy Giuliani introduced his own plan for health care reform. It was not, as he freely admitted, a plan designed to cover everybody. Even after implementation, tens of millions would still lack insurance. But the most remarkable and telling thing about Giuliani's plan was the way he presented it. In the speech unveiling his plan, he spent no more than a third of the time talking about his policy ideas. For the rest of the time, he held forth on a different, more familiar topic: the evils of "socialized medicine."

This, too, was a harbinger—a harbinger of just how determined and dishonest the opposition to universal coverage would continue to be. Giuliani warned that universal coverage would mean queuing up in long lines, apparently oblivious to the fact that countries like France, Germany, and Switzerland don't have chronic waits for treatment. He also spoke of technology shortages endemic to universal coverage—even though the country with the most CT and MRI scanners, per person, is not the United States but yet another country with universal coverage: Japan. Probably the most memorable argument that Giuliani made was about cancer. Giuliani, a prostate cancer survivor, warned that he might have died if he lived in another country, since statistics show the United States is far superior to Europe when it comes to curing cancer.

It was a powerful argument, given Giuliani's history, but not such an accurate one. As public health experts were quick to note, America's ultra-high survival rate for prostate cancer reflected the fact that its physicians screened so aggressively for it, inevitably catching—and then treating—a great many tumors that posed no grave threat. (That hardly fit the definition of better health care, since the side effects of prostate cancer treatment include chronic incontinence and impotence.) As it happens, studies released in 2007 suggested that—after adjusting for that anomaly—overall survival rates in the U.S. were only slightly better, if at all, than the survival rates in countries like France and Switzerland, which (like the U.S.) tended to purchase the latest, most expensive drugs. As a matter of fact, the newest data showed that the U.S. survival rates for a few cancers were downright mediocre. It turned out that a patient diagnosed with stomach cancer had better odds in Germany or the Netherlands, among other European countries, while cervical cancer patients were more likely to survive in Belgium, Italy, the Netherlands, Switzerland, and most of Scandinavia.

Other new studies painted an even dimmer picture. While the United States was decidedly good at some things (such as treating breast cancer), it was decidedly bad at other things (such as kidney transplants). Some critics of American health care have used information like this to argue that U.S. health care was flat-out inferior to that of most European nations, a charge that probably went too far. In all fairness, there seemed to be at least some areas where, because of America's tendency towards intensive treatment, the U.S. system did seem to be the world's best—at least if you defined "best" as providing heroic treatments for seemingly lost causes. (Neonatal care seemed to be one such category.) But taking into account all of the available evidence, it was impossible to argue credibly that the U.S. was clearly superior to other countries—or that, by adopting universal health insurance, the rest of the developed world had somehow consigned its citizens to second-class medical care. As the conservative writer Ramesh Ponnuru conceded in an article for the *National Review*, "[T]he best national health-insurance programs do not bear out the horror stories that conservatives like to tell about them."

As universal coverage started picking up more support in late 2007—including from some prominent conservatives in Congress, such as Senators Bob Bennett of Utah and Charles Grassley of Iowa—the opponents began concentrating on yet a second theme: arguing that the problem of the uninsured wasn't as bad as it seemed. They said the government estimates on the uninsured were flawed, that many uninsured people were already eligible for programs like Medicaid, and that at least some of the uninsured were, in effect, uninsured by choice—going without coverage just because they preferred to spend their money elsewhere.

While these arguments held some element of truth, they obscured more than they revealed. The census estimates on the uninsured, like all such estimates, were indeed imprecise. But various studies, conducted by different researchers, had yielded a pretty consistent picture: At any one time, a very large number of Americans—somewhere around 45 million—had no insurance. Although many would not be uninsured the entire year, that was part of the problem: Security was fleeting, since people moved in and out of coverage. (Studies had shown that, over a two-year period, 80 to 85 million Americans lost insurance at some point—which is a complicated way of saying that more than a quarter of the population had at least some period in which they were without coverage, exposing them to all sorts of medical and financial problems.)

It was true that people eligible for Medicaid sometimes didn't participate. But, all too frequently, that was because they couldn't enroll—and then stay enrolled—because it was difficult to manage the cumbersome registration process while constantly floating in and out of eligibility, as working-class people with unstable incomes frequently did. And while some of the uninsured could indeed afford coverage, that wasn't an argument against universal coverage. It was an argument for it. Many of these people would end up in emergency rooms someday, relying upon charity care—the cost of which doctors and hospitals would eventually pass along to taxpayers and people paying private insurance premiums. A universal system would force these people to pay into the system from the get-go, forcing them to take responsibility and bear part of the financial burden for the mutual (if not entirely adequate) protection they would ultimately enjoy. This was a big reason why some some self-described conservatives, such as Romney and Schwarzenegger, supported universal health care, even though Romney—in his efforts to capture the Republican presidential nomination—spent much of 2007 distancing himself from his own plan.

Still, the fact that critics of universal coverage were using misleading statistics didn't mean they couldn't still win the political battle to come. The strategy of equating universal coverage with inferior socialized medicine had a long history of success, going back to the 1940s, when the AMA used that argument to derail Harry Truman's proposal, and even back to the early New Deal, when medical societies and their allies tried to stymie nascent universal coverage efforts there. As for the strategy of disputing the number of uninsured, that fed right into the most dangerous myth of all: the belief of so many middle-class Americans that losing access to affordable medical care was somebody else's problem.

The Frost family of Baltimore knew better—just like Gary Rotzler, Janice Ramsey, Elizabeth Hilsabeck, and all the other people from *Sick* who'd learned this lesson the hard way. One by one, they'd discovered that having a home, a job, or even a college degree didn't necessarily mean you could get the health care you needed—or that you'd be protected from financial catastrophe in the event of a medical crisis. And, with each passing day, more people with more stories like these came forward. Was the rest of America paying attention? Was it possible this health care debate would turn out differently than the previous ones? Only time would tell, but for the first time in a long time, it was possible to be a little optimistic.

SOURCES AND NOTES

This is a work of nonfiction, based on original reporting and research over the course of five years. The names, places, and events are real. My primary sources for the narrative sections were the featured protagonists and other people with direct knowledge of their experiences. When a statement appears in quotation marks, it is because at least one source remembered hearing those words. When I describe a character's thoughts, it is because the character told me what he or she remembered thinking. Given the imperfections of human recollection, I have tried diligently to confirm what my interview subjects told me. The notes in this section provide insight into how I sourced and corroborated key narrative facts, drawing not only on interviews but also on correspondence, medical files, and other forms of documentation.

In describing the workings of the health care industry and public policy, I have drawn on hundreds of interviews—some conducted exclusively for this book, others conducted for articles that I was writing at the time. The notes detail those instances in which an interview produced a specific fact. Here, in addition to the people mentioned there, is a list of sources who provided background information: Henry Aaron, Drew Altman, Marcia Anderson, Stuart Atlman, Joe Antos, Bruce Bagley, Robert Ball, Tom Banning, Nina Bannister, Sarah Bianchi, Douglas Blayney, Robert Boorstin, Robert Brand, Mollyann Brodie, Andrew Brontman, Luke Brown, Esteban Burchard, Stuart Butler, Alwyn Cassil, Hobson Carroll, Joyce Clifford, Deborah Chollet, Forest Claypool, Peter Cunningham, Debbie Curtis, Doug Danziger, Helen Darling, Brian Davidson, Nancy-Ann DeParle, Ed

Domansky, David Dranove, Jan Emerson, John Emling, Bill Erwin, Judith Feder, Dennis Fitzgibbons, Grace Floutsis, Steve Foster, Elizabeth Fowler, Larry Gage, Thomas Garthwaite, Joseph Geevarghese, Tom Geoghegan, Paul Ginsburg, Mahit Ghose, Marsha Gold, Warren Gold, Irma Gonzales, John Goodman, Suzanne Gordon, Robert Graham, Michael Grassi, Patricia Gray, Tina Hanson-Turton, Howard Hiatt, Steve Hitov, Jesse Hixson, Catherine Hoffman, Jill Horwitz, Ed Howard, William Hsiao, Jeremy Hurst, Charles Idelson, Sylvia Drew Ivie, Roger Jenkins, Chris Jennings, Vicki Johnson, Rendy Jones, William Jones, Julius Kaplan, Gerald Katz, William Kelly, Michael Knotek, Jerome Lackner, Gaetan LaFortune, Larry Levitt, Jim Lott, Timothy McCurry, Catherine McLaughlin, Leon Malmud, Cindy Mann, Al Mansfield, Barbara Masters, Burt Margolin, Grace Marie-Turner, Jim Martin, Glen Maxey, Harris Meyer, Megan McAndrew, John McDonough, Lisa McGiffert, Torre Muhammad, Elliot Naishtat, Ed Newton, David Nexon, Len Nichols, Rebecca Owens, Edwin Park, Donald Parks, Mark Pauly, Thomas O'Toole, Linda Peeno, Zaida Perez, Ron Pollack, James Quiggle, Mitch Rabkin, Ken Robbins, Victor Rodwin, Dean Rosen, Neil Rosenberg, Michael Rosenblatt, David Rousseau, Diane Rowland, Alan Sager, Greg Scandlen, Andy Schneider, Scott Selco, Tammy Seltzer, William Shernoff, Frank Sloan, Andy Stern, Ann Thornburg, Kenneth Thorpe, Janet Trautwein, Lynn Trimble, Peter Van Vranken, Rick Wade, Judith Waxman, Bryant Welch, Tim Westmoreland, Pam White, Gail Wilensky, Arnold Widen, Lucian Wulsin, Anthony Wright, Quentin Young, and Stephen Zuckerman. A few additional sources appear in the acknowledgments. Absent from either list are sources who spoke to me on the condition I would not use their names—mostly health care industry insiders and hospital employees who said anonymity would allow them to speak with greater candor, either about the nature of their businesses or specific stories in the narratives.

In reconstructing the story of American health care during the twentieth century, I used several historical benchmarks: the original reports of the Committee on the Costs of Medical Care, published during the late 1920s and early 1930s; *Doctors, Patients, and Health Insurance*, an exhaustive survey that Herman M. Somers and Anne R. Somers produced in 1961, under the auspices of the Brookings Institution; annual reports from the Health Insurance Association of America (which recently merged with another trade group and renamed itself America's Health Insurance Plans), dating back to the 1940s; and assorted government reports.

My other source for history was the unofficial canon of great works on health care (plus a few lesser-known volumes that perhaps deserve wider attention): *Health against Wealth* by George Anders; *The System* by David Broder and Haynes Johnson; *The Blues: A History of the Blue Cross and Blue Shield System* by Robert Cunningham III and Robert Cunningham Jr.; *HMOs and the Politics of Health System Reform* by Joseph L. Falkson; *The Divided Welfare State* and *The Road to Nowhere* by Jacob Hacker; *Free for All? Lessons from the RAND Health Insurance Experiment* by Joseph P. Newhouse; *The Political Life of Medicare* by Jonathan Oberlander; *The Politics of Medicare* by Theodore J. Marmor; *The Care of Strangers* by Charles E. Rosenberg; *The Social Transformation of American Medicine* by Paul Starr; *In Sickness and in Wealth* by Rosemary Stevens; and *Welfare Medicine in America* by Robert Stevens and Rosemary Stevens.

Particularly where I discuss public policy and the present workings of the health care industry, I have consulted academic literature, relying primarily on work published in such peer-reviewed journals as the *New England Journal of Medicine*; *Health Affairs*; and the *Journal of Health Policy, Politics, and the Law*. I also relied heavily on analysis from two well-respected, nonpartisan sources—the Center for Studying Health Systems Change and the Henry J. Kaiser Family Foundation, which also underwrote a two-year fellowship that allowed me to work on this book. In general, I have tried to avoid using work published by interest groups, trade organizations, and other researchers that might not conform to formal academic standards. (The exceptions are those cases where such work was the only information available.)

I owe much of my education about health care—and, in several cases, leads to characters for the book—to the work of fellow journalists. While the periodical references in my footnotes are testimony enough to this, I want to mention a few whose articles were particularly helpful. Milt Freudenheim and Robert Pear of the *New York Times*, along with Julie Rovner of *National Journal* and National Public Radio, for their coverage of health policy; Peter Gosselin, of the *Los Angeles Times*, for his articles on economics; Lucette Lagnado, of the *Wall Street Journal*, for her stories on hospitals' debt collection; Ellen R. Schultz, also of the *Journal*, for her articles on pensions and retirement benefits; Charles Ornstein and Tracy Weber, also of the *Los Angeles Times,* for their stories on Kaiser Permanente and the Los Angeles County health system; and Larry Tye of the *Boston Globe,* for his coverage of crowding in emergency rooms. Another invaluable source was John Iglehart's exhaustive series of articles on the U.S. health care system, which appeared in the *New England Journal of Medicine* during the 1980s and 1990s.

Most of the originally reported material in this book is appearing for the first time here. The notable exception is Chapter 6, which I adapted from an article that I wrote for the *New York Times Magazine* in 2004. In addition, analytical sections throughout the book draw on, and overlap with, my past writings for the *New Republic.* Rather than list each source for those stories, I have simply noted the articles themselves.

ABBREVIATIONS:

BG—*Boston Globe*

LAT—*Los Angeles Times*

NYT—*New York Times*

WP—*Washington Post*

WSJ—*Wall Street Journal*

BW—*Business Week*

EBRI—Employee Benefit Research Institute

KFF—Henry J. Kaiser Family Foundation

HSC—Center for Studying Health System Change

GAO—U.S. Government Accounting Office
(renamed the Government Accountability Office in 2004)

MEDPAC—Medicare Payment Advisory Commission

HA—*Health Affairs*

JAMA—*Journal of the American Medical Association*

NEJM—*New England Journal of Medicine*

TNR—*The New Republic*

INTRODUCTION: BOSTON

p.ix It was 4:43 on a clear November afternoon . . . Bill Mergendahl and Keri Powers, interviews with author (Mergendahl was the dispatcher for Professional Ambulance Service in Cambridge; Powers was one of the paramedics on the scene); Massachusetts Department of Public Health Investigative Report, January 2001; and Larry Tye, "For One Patient, Delay in Care Proved Fatal," *BG*, December 25, 2000.

p.xii Still, as one source familiar with the case . . . Tye, "Patient's Death Draws Review by State Agency," *BG*, January 3, 2001.

p.xii . . . the overcrowding epidemic was routinely jeopardizing the well-being of patients . . . Peter Gosselin, "Amid Nationwide Prosperity, ERs See a Growing Emergency," *LAT*, August 6, 2001; and Catherine M. Dunham, "Emergency Department Overcrowding in Massachsuetts: Making Room in Our Hospitals," Issue Brief for the Massachusetts Health Policy Forum, 2001.

p.xiii Nancy Ridley had her own troubles in the ER . . . Tye, "ER Crisis Hits Close to Home for Officials," *BG*, May 21, 2001.

p.xiii In Atlanta, an ambulance crew carrying a patient . . . Andy Miller, "Too-Crowded Hospitals Play Diverting Game," *Atlanta Journal-Constitution*, June 15, 2001.

p.xiii When the mother of a forty-year-old Cleveland man . . . Diane Solov and Regina McEnery, "Hospital Ambulance Diversions Keep Climbing," *Plain-Dealer*, May 6, 2001.

p.xiii In suburban Houston, when a twenty-one-year-old man . . . Guy Clifton, interview with author; Mary Ann Roser, "In Critical Condition: Diverted Ambulances Reflect 'Looming Crisis' for Austin Hospitals," *Austin American-Statesman*, January 20, 2002.

p.xiv Fed up with incidents like that . . . Guy Clifton, interview with author.

p.xiv . . . a symptom of other systemic problems now plaguing medical care . . . Linda R. Brewster, Liza S. Rudell, and Cara S. Lesser, "Emergency Room Diversions: A Symptom of Hospitals under Stress," HSC, Issue Brief No. 38, May 2001. For details on the role these factors played in emergency room overcrowding in Massachusetts, see Dunham, "Emergency Department Overcrowding"; Katherine Lutz, "Mental Patients Turning to ERs: Cuts Force Many to Try Hospitals," *BG*, August 20, 2003; Raja Mishra, "ERs Say No When Beds Are Scarce; in Many Cases, Care Is Delayed," *BG*, December 10, 2001; Larry Tye, "Emergency Room Crisis Worsening; Ambulances in Boston Area Often Diverted," *BG*, December 25, 2000; Tye, "ER Diversions Rose Again at Year's End," *BG*, January 6, 2001.

p.xv Under Clinton's proposal, the government would have ... See, generally, David S. Broder and Haynes Johnson, *The System: The American Way of Politics at the Breaking Point* (Boston, Mass.: Little, Brown, 1996); Theda Skocpol, *Boomerang: Clinton's Health Security Effort and the Turn against Government in U.S. Politics* (New York: Norton, 1996); Paul Starr, "What Happened to Health Care Reform?" *American Prospect*, Winter 1995, pp. 20–31.

p.xv Most Americans, after all, still had health insurance ... Robert J. Blendon, correspondence with author; Blendon, Mollyann Brodie, and John Benson, "What Happened to Americans' Support for the Clinton Health Plan," *HA*, Vol. 14, No. 2, Summer 1995, pp. 7–23. See also Judi Hasson, "Poll: Sense of Urgency Fades, Cost Worries Grow," *USA Today*, June 30, 1994; Michael L. Millenson, "Middle Class Turns a Cautious Eye on Health-Care Reform," *Chicago Tribune*, January 31, 1994; William Schneider, "A Fatal Flaw in Clinton's Health Plan?" *National Journal*, November 6, 1993, p. 2696; *Time*/CNN Poll, by Yankelovich Partners, February 10, 1994. As Schneider noted, "[O]nly a fourth of Americans say that Clinton's plan would make them feel more secure. A third say they would be less secure. That's supposed to be the big selling point of this plan, and it isn't selling."

p.xv "I've got pretty good health care ... "Rep. Newt Gingrich Takes Calls on Health Care Reform," CNN, August 8, 1994.

CHAPTER 1: GILBERTSVILLE

p.1 New York's Leatherstocking Country sits at the northern foothills ... James Fenimore Cooper, *Pioneers* (New York: Penguin Putnam, 1985), pp. 13–14.

p.2 Perhaps nowhere is this more evident than in the village of Gilbertsville. ... Descriptions of Gilbertsville and its population come from my visit there in 2004, plus interviews with several residents, including the municipal historian and archivist Leigh Eckmair. For historic photographs and other background on Gilbertsville and the surrounding area, I consulted the personal online journal of Philip Lord, Jr., at http://home.att.net/~_lord2/butternut.html.

p.3 ... Gary and Betsy Rotzler moved to Gilbertsville ... Gary Rotzler, Mary Lou Harvey, and Patricia Stevenson, interviews with author.

p.4 ... the erosion of job-based insurance ... U.S. Census Bureau, *Income, Poverty, and Health Insurance Coverage in the United States: 2003* (Washington: U.S. Commerce Department, 2004), p. 53. Note that the figure was rising as both a raw number and a proportion of the population.

p.4 . . . the majority of these newly uninsured were neither destitute nor truly jobless . . . KFF, *The Uninsured: A Primer—Key Facts about People without Health Insurance*, December 2003, p. 4.

p.5 . . . historians generally trace its development back to ancient Babylonian traders . . . Peter L. Bernstein, *Against the Gods: The Remarkable Story of Risk* (New York: Wiley, 1998); C. G. Lewin, *Pensions and Insurance before 1800: A Social History* (East Linton, Scotland: Tuckwell, 2003), pp. 6–16, 336, 420, and 428; George Rosen, *A History of Public Health*, expanded edition with an introduction by Elizabeth Fee (Baltimore, Md.: Johns Hopkins University Press, 1993), pp. 417–420.

p.5 America's first private insurance company . . . Nicholas B. Wainright, "Philadelphia's Eighteenth-Century Fire Insurance Companies," *Transactions of the American Philosophical Society*, New Ser., Vol. 43, No. 1, 1953, pp. 247–248. Some would argue that Franklin's wasn't really the first, since a scheme to insure homes against fire had appeared in Charleston, South Carolina, in 1735—fifteen years before Franklin got to work on his plan in Philadelphia. But the plan in Charleston lasted just five years before it was shuttered.

p.5 But it was not until the early twentieth century that the idea of using insurance . . . For the early history of health insurance in America, I relied primarily on a few key sources: Jacob S. Hacker, *The Divided Welfare State: The Battle over Public and Private Social Benefits in the United States* (New York: Cambridge University Press, 2002), pp. 179–220; Institute of Medicine, *Employment and Health Benefits: A Connection at Risk* (Washington, D.C.: National Academy Press, 1993), pp. 49–76; Jill Quadagno, *One Nation Uninsured: Why The U.S. Has No Health Insurance* (New York: Oxford University Press, 2005), pp. 17–76; Paul Starr, *The Social Transformation of American Medicine: The Rise of a Sovereign Profession and the Making of a Vast Industry* (New York: Basic Books, 1982), pp. 235–279; and Melissa A. Thomasson, "From Sickness to Health: The Twentieth-Century Development of U.S. Health Insurance," *Explorations in Economic History*, Vol. 39, 2002, pp. 233–253.

p.6 As one scientist of the era famously remarked . . . Lawrence J. Henderson, quoted in Alan Gregg, *Challenges to Contemporary Medicine* (New York, 1956), p. 13. (Source: Herbert E. Klarman, "Medical Care Costs and Voluntary Health Insurance," *Journal of Insurance*, Vol. 24, No. 1, September 1957, p. 26.)

p.6 . . . the average cost of a week in the hospital began to exceed . . . Michael M. Davis, "The Committee on Costs of Medical Care Makes Its Report," *Modern Hospital*, Vol. 39, No. 6, December, 1932, p. 43.

p.6 "Very few of these families are indigent . . . Louis S. Reed, "The Ability to Pay for Medical Care," *Journal of the American Statistical Association*, Vol. 28,

No. 181, Supplement: Proceedings of the American Statistical Association, March 1933, p. 107.

p.6 . . . the Committee on the Costs of Medical Care, a blue-ribbon commission . . . Committee on the Costs of Medical Care, *Medical Care for the American People; the Final Report of the Committee on the Costs of Medical Care, Adopted October 31, 1932* (Chicago, Ill.: University of Chicago Press, 1932). See also Davis, pp. 41–46; C. Rufus Rorem, "The Costs of Medical Care," *Accounting Review*, Vol. 7, No. 1, March 1932, p. 38; and Starr, pp. 261–266.

p.7 One of those institutions was Baylor Hospital . . . Robert Cunningham III and Robert M. Cunningham Jr., *The Blues: A History of the Blue Cross and Blue Shield System* (DeKalb: Northern Illinois University Press, 1997), pp. 3–33. The quotation about Baylor's debt comes from Lana Henderson, *Baylor University Medical Center, Yesterday, Today, and Tomorrow*, p. 71, cited in Cunningham and Cunningham, p. 4. Note that Kimball had not invented the idea of using the workplace as a basis for providing medical care. During the nineteenth century, railroad and mining companies had gotten into the habit of giving employees access to a company doctor or clinic, sometimes in exchange for a small payroll deduction. In 1917, textile laborers in North Carolina had pooled workers' donations to help pay for medical care from a local hospital. A welfare association in Connecticut later did the same.

p.8 A habit was forming . . . Herman Miles Somers and Anne Ramsey Somers, *Doctors, Patients, and Health Insurance: The Organization and Financing of Medical Care* (Washington, D.C.: Brookings Institution, 1961), p.263; Hacker, 212-219. Labor's relative ambivalence about universal health care would also prove pivotal during Clinton's fight over health care, although it had a different genesis: labor leaders were preoccupied fighting the North American Free Trade Agreement (NAFTA), another of Clinton's initiatives. By the time that fight was over and labor, though still angry over NAFTA's enactment, was ready to put its considerable political muscle behind Clintoncare, the plan was for all intents and purposes dead politically. Ira Magaziner, interview with author. (Magaziner was senior adviser to the president for policy development and led the Clinton administration's health care task force.) See also Theda Skocpol, *Boomerang: Clinton's Health Security Effort and the Turn against Government in U.S. Politics* (New York, Norton, 1996), pp.91-92.

p.9 By 1980, most full-time workers at large companies . . . See EBRI, *EBRI Databook on Employee Benefits*, November 1995. Online at http://www.ebri.org/pdf/publications/books/databook/DB.Chapter%2004.pdf.

p.10 . . . health insurance premiums were eating into their paychecks . . . Most economists believe that the money businesses spend on health insurance for their employees is, for the most part, money that would otherwise have gone into their employees' paychecks. But it's not clear whether that's always true in specific cases—whether, for example, individual workers or simply groups of workers on average feel the impact. See Linda Blumberg, "Who Pays for Employer-Sponsored Health Insurance?," *HA*, Vol. 18, No. 6, November/December 1999, pp. 58–61.

p.10 But that neat arrangement started to break down . . . In 1991, just 11.5 percent of manufacturing workers were uninsured, compared with 25.3 percent of retail workers and 29.5 of workers in "personal/entertainment services." Deborah Chollet, Employer-Based Insurance in a Changing Job Market," *HA*, Vol. 13, No. 1, Spring 1994, pp. 315–326. Chollet notes that manufacturers faced the equivalent of a double burden relative to other businesses. Not only were they more likely to insure their own workers; they were more likely to insure dependents—who might otherwise have demanded coverage from other businesses. On the whole, she found, manufacturers were in effect subsidizing the insurance for workers in the retail and service industries.

p.10 Just as cheap health insurance premiums and tight labor markets once created . . . When two scholars from the Urban Institute built a mathematical model to track the decline in employer-sponsored health insurance and project future changes, they found that "the single most important detrimental trend affecting offer and take-up [of health insurance] is the rising share of part-time workers in the labor force. Part-time workers are far less likely than full-time workers are to be offered employer coverage, and their often modest incomes make the employee share of any offered coverage less affordable." Glenda Acs and Linda J. Blumberg, "How a Changing Workforce Affects Employer-Sponsored Health Insurance," *HA*, Vol. 20, Number 1, 2001, p. 178. For more on the way the reengineering of company workforces eroded job-based health coverage, see Henry S. Farber and Helen Levy, "Recent Trends in Employer-sponsored Health Insurance Coverage: Are Bad Jobs Getting Worse?" *Journal of Health Economics*, Vol. 19, 2000, pp. 93–119.

p.10 As of 2005 less than half of Wal-Mart's employees . . . *Kaiser/HRET Survey of Employer-Sponsored Health Benefits* (Washington, D.C.: KFF, 2004); Wal-Mart public relations office, interview with Elspeth Reeve on behalf of author; and Wal-Mart Watch, "Background on Wal-Mart Health Insurance Coverage," online at http://walmartwatch.com/home/pages/healthcare#background.

p.10 Those companies unable to scale back their commitment . . . James Bennet, "Auto Talks Hang on Health Costs but Workers Are Loath to Chip In," *NYT*, September 6, 1993; and Ed Gartsten, "GM Health Care Bill Tops $60 Million," *Detroit News*, March 11, 2004. One reason for the sharp rise: as with many older manufacturing companies, a large portion of GM's medical expenses are benefits to retirees who were promised generous coverage and, now that they've reached old age, are consuming large amounts of it. For more about retirement health benefits, see Chapter 4.

p.11 . . . Gary's employer—Bendix—had attempted to take over . . . John S. De-Mott, "White Knights and Black Eyes," *Time*, February 14, 1983, p. 56.

p.11 . . . Bendix had a reputation as one of central New York's best employers . . . L. Eckmair, Mary Crim, Shirley Musson, and G. Rotzler, interviews with author.

p.11 Gary figured he could either get out or be tossed out . . . G. Rotzler, interviews with author. Regarding the dismissal from ShopVac, Gary says that no reason was given.

p.12 . . . he was designated a temporary worker . . . Amphenol could not confirm the details of Gary's employment. However, a company official did confirm that, like most companies, Amphenol sometimes laid off workers during lean times and then, when business picked up, hired them back as independent contractors because that arrangement was less costly. Sheri Youngs, interview with author. (Youngs was coordinator of staffing and compensation for Amphenol in 2006.)

p.13 The change was particularly evident to Betsy's younger sister . . . P. Stevenson, interviews with author.

p.14 Betsy's younger brother, Steve, picked up on the change . . . S. Harvey, interview with author.

p.14 At first, Betsy's family assumed she was just reacting . . . M. Harvey, S. Harvey, P. Stevenson, interviews with author.

p.15 . . . she was frequently shy with outsiders . . . M. Crim, L. Eckmair, S. Musson, Jean Postighone, interviews with author.

p.16 Trisha, on the other hand, had no such inhibitions. . . . M. Harvey, G. Rotzler, P. Stevenson, interviews with author. It was impossible to verify the story about Betsy's efforts to see her former obstetrician, as Trisha was the only person who recalled the story, and the doctor no longer worked in the area. However, in the United States it is common for both physicians and hospitals to refuse non-emergency services to uninsured patients unless they are able to pay right away.

p.16 "Every time the children got sick . . . M. Harvey, interviews with author.

p.16 So instead of getting professional medical help . . . M. Harvey, G. Rotzler, P. Stevenson, interviews with author.

p.17 . . . the U.S. health care system has been enormously successful . . . David M. Cutler, *Your Money or Your Life: Strong Medicine for America's Health Care System* (New York: Oxford University Press, 2004), pp. 80–85.

p.17 Magnetic resonance imaging (MRI) . . . Howard Markel, correspondence with author. (Markel is a professor of pediatrics and the history of medicine, University of Michigan Medical School.)

p.17 . . . while most Americans assume that people without health insurance still get good health care, a mountain of data suggests otherwise . . . Institute of Medicine, *Care without Coverage: Too Little, Too Late* (Washington, D.C.: National Academy Press, 2002), pp. 3–5 and 47–72. For a detailed discussion about how some screening technologies may be misused or overused, see Shannon Brownlee, "Search and Destroy: Why Mammograms Are Not the Answer," *TNR*, April 22, 2004, p. 24.

p.17 Betsy was, in this sense, entirely typical . . . M. Harvey, A. Rotzler, G. Rotzler, P. Stevenson, interviews with author.

p.20 Gary didn't want to contemplate finances at this point . . . G. Rotzler, interviews with author.

p.20 The undertakers hadn't applied much makeup . . . P. Stevenson, interviews with author.

p.21 When he first learned that Betsy might have cancer . . . G. Rotzler, interviews with author.

p.21 As word of Betsy's death . . . M. Crim, L. Eckmair, S. Musson J. Postighone, G. Rotzler, interviews with author.

p.21 Such financial devastation was not unusual . . . Warren's research provoked a spirited debate among academics. Critics suggested that Warren and her sometime collaborators, David Himmelstein and Steffie Woolhandler, had overstated the influence of medical debt. But even the critics conceded that medical debt was a major component of financial hardship for a large number of Americans, including some who actually had insurance. For Warren's findings, see Himmelstein, Warren, Deborah Thorne, and Woolhandler, "MarketWatch: Illness and Injury as Contributors to Bankruptcy," *HA* Web Exclusive, February 2, 2005; Himmelstein, Warren, Thorne, and Woolhandler, "Discounting the Debtors Will Not Make Medical Bankruptcy Disappear," *HA*, Vol. 25, No. 2, March/April 2006, pp. w84–w88. For the critique, see David Dranove and Michael L. Millenson, "Medical Bankruptcy: Myth versus Fact," *HA*, Vol. 25, No. 2, March/April 2006, pp. w74–w83; Dranove and Millenson, "Medical Bankruptcy: Dranove and Millenson Respond," *HA*, Vol. 25, No. 2, March/April 2006, p. w93.

p.22 "Obviously, I'm sure she was just hoping . . . G. Rotzler, interviews with author.

p.22 Trisha suspected that Betsy was probably too embarrassed . . . P. Stevenson, interviews with author.

p.22 It was all such a shame . . . L. Eckmair, Mary Crim, and Shirley Musson, interviews with author.

p.22 . . . doctors and hospitals will treat anybody with an emergency . . . The law, the Emergency Medical Treatment and Labor Act (EMTALA), prohibits health care providers from turning away emergency patients because they cannot pay. Note, however, that it does not prohibit doctors or hospitals from charging for emergency services after the fact; this is why so many uninsured people end up facing large debts. For more on EMTALA and its effect on hospitals' behavior, see Chapter 6.

p.22 "In this country, we do not see many people dying in the streets . . . Institute of Medicine, *Care without Coverage: Too Little, Too Late* (Washington, D.C.: National Academy Press, 2002), p. ix.

p.22 A patient diagnosed with stage-zero breast cancer has a 95 percent chance . . . Carol Rosenberg, M.D., interview with author. Dr. Rosenberg is associate professor of medicine at Boston University Medical School and a hematologist-oncologist who specializes in breast cancer.

p.23 According to one study originally published in the *New England Journal of Medicine* . . . John Z. Ayanian, Betsy A. Kohler, Toshi Abe, and Arnold M. Epstein, "The Relation between Health Insurance Coverage and Clinical Outcomes among Women with Breast Cancer," *NEJM*, 1993, pp. 326–331. Cited in Institute of Medicine, *Care without Coverage*, p. 54. The report cites dozens of other peer-reviewed studies reaching essentially the same conclusion about breast cancer and other cancers.

p.23 Uninsured Americans with a chronic illness . . . Institute of Medicine, *Care without Coverage*, pp. 1–16. Most of the researchers cited in the report said that they controlled for variables such as economic status and specific disease.

p.23 . . . nobody can ever know for certain . . . Once the cancer spread to her spine, it would have been considered terminal. But, even by that point, it's possible therapy could have prolonged her life.

p.23 Closure for Gary came slowly. . . . G. Rotzler, interviews with author; G. Rotzler, "Raw" and "Why Not My Love," unpublished poems posted at www.widownet.org.

p.24 Betsy's mother, Mary Lou, channeled her feelings . . . M. Harvey, interviews with author.

p.24 Gary, who had stopped voting Republican in the 1990s . . . The interviews for this book were the first time Gary talked publicly about Betsy's story. He said that one reason he was reluctant to join Mary Lou in becoming a spokesman for the uninsured was that he feared retribution from his employer while his children still depended on him for health insurance. At the time of the interviews, his youngest son, Luke, had recently turned seventeen and was bound for college. G. Rotzler, interviews with author.

p.24 The ranks of the uninsured had, briefly, stopped growing in the late 1990s . . . James D. Reschovsky, Bradley C. Strunk, and Paul Ginsburg, "Why Employer-Sponsored Insurance Coverage Changed, 1997–2003," HA, Vol. 25, No. 3, May/June 2006, pp. 774–782.

p.24 . . . Democrats seemed to be shaking off the political hangover . . . Spokespersons and policy directors for the leading Democratic presidential candidates of 2004, interviews with author. ("Checkups: Rating the Dems on Health Care," TNR, June 9, 2003.)

CHAPTER 2: DELTONA

p.27 For nearly two years, Janice Ramsey . . . Janice Ramsey, interview with author; Florida Department of Insurance, *Immediate Final Order in the Matter of American Benefit Plans, et al*, Case No. 43161-02-CO; KFF, *Briefing: Falling through the Cracks: Stories of How Health Insurance Can Fail People with Diabetes*, (Washington, D.C.) February 8, 2005.

p.30 In a period of roughly two years, thousands of people . . . Mila Kofman, Kevin Lucia, and Eliza Bangit, *Proliferation of Phony Health Insurance: States and the Federal Government Respond* (Washington, D.C.: Bureau of National Affairs, 2003), p. 14.

p.31 Although the Blues were private entities . . . Robert Cunningham III and Robert M. Cunningham Jr., *The Blues: A History of the Blue Cross and Blue Shield System* (DeKalb, Illinois: Northern Illinois University Press, 1997), pp. 3–33; Institute of Medicine, *Employment and Health Benefits: A Connection at Risk* (Washington: National Academy Press, 1993), pp. 66–69; Rosemary Stevens, *In Sickness and Wealth*, 2nd edition, (Baltimore, Md.: Johns Hopkins University Press, 1999), pp. 171–199.

p.32 The "brakes Blue Cross has applied to our swing towards socialization" . . . "Blue Cross Is Called Socialization Brake," NYT, May 27, 1950.

p.32. . . more than 20 million people had enrolled in Blue Cross . . . "20,000,000 in Blue Cross: John R. Mannix Predicts Full National Coverage," NYT, January 14, 1946.

p.32. . . commercial insurers had at various times offered something called "sickness" benefits . . . Edwin L. Bartleson et al. for the Society of Actuaries, *Health Insurance Provided through Individual Policies* (Baltimore, Md.: Waverly, 1963), pp. 1–2; Herman Miles Somers and Anne Ramsay Somers, *Doctors, Patients, and Health Insurance* (Washington, D.C.: Brookings Institution, 1961), pp. 261–262.

p.33 . . . they were a nightmare to manage financially . . . The carriers' financial troubles after the stock crash of 1929 had something to do with the decision, too.

p.33 Their arrival signaled a critical change, however. . . . Hacker, pp. 202–206; Institute of Medicine, pp. 71–72; Melissa A. Thomasson, "Early Evidence of an Adverse Selection Death Spiral? The Case of Blue Cross and Blue Shield," *Explorations in Economic History*, 41 (2004), pp. 313–328.

p.34 . . . life insurers had since replaced casualty insurers . . . In 1958, life insurance companies were responsible for 88 percent of the commercial health insurance business. Somers and Somers, p. 263.

p.34 Among the more bizarre factors insurers sometimes considered was height . . . Somers and Somers, p. 276.

p.34 . . . the largest commercial insurers generally steered clear of the individual market . . . In 1958, of the top five commercial insurers offering group health coverage, two reported no individual coverage at all and only one, Prudential, was among the top five offering individual coverage. Somers and Somers, p. 263.

p.34 . . . the insurance industry made no excuses for its general approach . . . In breaking up applicants into different risk classes, the Society of Actuaries explained in one of its guidebooks, "within practical limits, rating classes should be sufficiently narrow that one section of the class does not subsidize the other materially." Bartleson, p. 106. See also Somers and Somers, pp. 274–276. At one point, the head director of information and research of the Health Insurance Association of America argued that community rating was anticompetitive, encouraged excessive use of medical care, and was contrary to "sound insurance principles." J. F. Follmann Jr., "Experience Rating vs. Community Rating," *Journal of Insurance*, Vol.29, No. 3 (September 1962), pp.402–415.

p.35 . . . the Blues finally began a slow retreat from their founding principles. . . . Dickerson, *Health Insurance*, 3rd ed., Irwin Series in Risk and Insurance (Homewood, Ill.: Irwin, 1968), pp. 328–329. By 1959, most Blue Cross plans were using experience rating partly or exclusively as their method of setting premiums for their group clients. Thomasson, "Early Evidence of an Adverse Selection Death Spiral?," pp. 317–318.

p.35 But the trend toward more commercial behavior continued. "Comparing Blue Cross and Blue Shield Plans with Commercial Insurers" (Washington, D.C., 1986).

p.35 . . . "the community rate concept is, for all practical purposes, dead." Odin W. Anderson, *The Uneasy Equilbrium: Private and Public Financing of Health Services in the United States* (New Haven, Conn.: College and University Press, 1968), p. 202.

p.35 "Any lingering hope that private insurance might provide . . ." R. Cunningham, correspondence with author; Jacob Hacker, *The Great Risk Shift* (New York: Oxford University Press, 2005), pp. 146–147; EBRI, "ERISA and Health Plans," Issue Brief Number 167, (Washington, D.C.), November 1995.

p.36 . . . Blue Cross is a wholly different enterprise . . . In order to convert to for-profit status, a Blue Cross plan had to win approval from state regulators and find some way to "pay back" the tax subsidies the plan had accumulated over the decades. Most of the plans did so by endowing nonprofit foundations dedicated to health care issues or helping to fund programs that subsidized coverage for the poor and uninsured. Many critics protested the sales, arguing that the sums were too small, given the enormous tax breaks the plans had once enjoyed. But such critics were able to block only a few proposed sales.

p.36 . . . many have adopted the same practices as the commercial carriers . . . Office of Technology Assessment (OTA), *Medical Testing and Health Insurance* (Washington, D.C., 1988), pp. 57–58, 78.

p.36 The smaller commercial carriers are even more selective. . . . OTA, 1988, pp. 55–80.

p.36 "As an underwriter, you must be aware and ask . . . The speaker was John Clark, risk selection director of individual health at Mutual/United of Omaha. "Medical Underwriting—A Retrospective," transcript from Society of Actuaries annual meeting in San Francisco, October 17–22, 1999. Published in *Record*, Vol. 25, No. 3, p. 8.

p.37 Insurers and others who defend these practices . . . Jonathan Shreve, FSA, "Medical Underwriting: Approaches and Regulatory Restrictions," Milliman Global, presentation. Date unknown.

p.38 Individually, diabetics can pay thousands of dollars a year in medical costs . . . Karen Pollitz et al., "Falling through the Cracks: Stories of How Health Insurance Can Fail People with Diabetes" (Washington, D.C.: American Diabetes Association and Georgetown University Health Policy Institute, February 8, 2005).

p.38 . . . insurers also know that beneficiaries change plans . . . Kathryn Klein, Sherry A. Glied, Danielle Ferry, "Entrances and Exits: Health Insurance Churning, 1998–2000," (New York: The Commonwealth Fund, September 2005).

p.39 Janice Ramsey had discovered this firsthand in 1999. . . . J. Ramsey, interviews with author.

p.40 . . . companies tend to deny outright between 15 and 25 percent of the applications . . . Mark Merlis, *Fundamentals of Underwriting in the Nongroup Health
Insurance Market: Access to Coverage and Options for Reform* (Washington,
D.C.: National Health Policy Forum, April 13, 2005), pp. 9–12. The National Association of Health Underwriters, among others, has suggested that those figures
present a jaundiced, unfair view, because in reality only 1 million or so Americans
are truly "uninsurable" and most people could find some kind of coverage if they
shopped around for long enough. But, in many cases, the only benefits available
are extremely expensive, or fail to cover certain medical conditions, or impose
long waiting times before full coverage, rendering them effectively worthless to
beneficiaries with serious, ongoing medical conditions.

p.40 Some of the most illustrative evidence . . . K. Pollitz, interview with author
and testimony before the U.S. Senate Committee on Health, Education, Labor,
and Pensions, March 12, 2002.

p.41 One day in the late 1940s . . . Joseph Ingraham, "'Gyps' in Insurance on
Health Fought," *NYT*, April 12, 1949.

p.42 About thirty years later . . . Kofman, Lucia, and Bangit, *Proliferation of
Phony Health Insurance*, p. 14.

p.43 . . . creating considerable misunderstanding over who would be responsible
. . . Kofman, Lucia, and Bangit, p. 17.

p.44 In addition to using promotional material that looked authentic . . . Bryan Grow,
"Health Insurance Scams That Will Make You Sick," *BW*, August 22, 2002, p. 40.

p.44 . . . they concentrated most of their efforts in the south . . . According to
a study by the Government Accounting Office, of the seven states with twenty-
five or more bogus insurers, five were in the south, with heavy concentrations in
Florida and Texas. GAO, *Private Health Insurance: Employers and Individuals
Continue to Face Challenges in Providing Coverage* (Washington D.C., October
2001). Opinions differ on exactly why these states were so vulnerable, but most
of the experts I consulted felt that the high proportion of self-employed people
and small businesses in these states (relative to other parts of the country) meant
the purchasers of health insurance were, on the whole, less sophisticated. They
might not have the same experience and expertise as benefit managers at large
companies; or, at the very least, they lacked the financial resources to hire good
consultants to do that job on their behalf. (Luke Brown, M. Kofman, and Pam
White, interviews with author. Brown was an attorney who, along with White,
worked for the state of Florida on insurance fraud.)

**p.44 The insurance plans sold most of their coverage to individuals through
licensed agents.** . . . Robert Brace, L. Brown, M. Kofman, P. White, interviews
with author. Brace was an attorney representing victims of fraud.

p.45 Even in those instances when officials were able to shut down the fake plans immediately . . . GAO, October, 2001.

p.45 . . . Richard Baer, the owner of a small greeting card shop . . . Richard Baer, M. Kofman, and P. White, interviews with author. See also Glenn Singer, "Florida Tightening Insurance Oversight," *Sun-Sentinel*, February 10, 2002; Harry Wessel, "Insurance Executives Plead Guilty," *Orlando Sentinel*, June 11, 2005.

p.46 . . . Rusty Baker had also purchased coverage from TRG. . . . Steven Meyer, interview with author. (Meyer was an attorney representing Cynthia Baker); Audrey Blackwell, "Scam Devastates Local Family," *Okeechobee News*, October, 1996.

p.46 Perhaps the most famous victim of an insurance scam in Florida was Pete Orr . . . Teri Orr, interview with author. See also Greg Groeller, "Health-Insurance Scams Hit Thousands of Floridians," *Orlando Sentinel*, April 14, 2002.

p.47 She was still alive, for one thing. . . . J. Ramsey, interviews with author.

p.47 . . . Janice heard about the Health Insurance Portability and Accountability Act . . . J. Ramsey, interviews with author; U.S. Department of Labor, "Frequently Asked Questions About Portability of Health Coverage and HIPAA" (viewed online at www.dol.gov/ebsa/faqs/faq_consumer_hipaa.html); GAO, "Private Health Insurance: Progress and Challenges in Implementing 1996 Federal Standards," May 1999.

p.48 Next Janice turned to the state of Florida . . . J. Ramsey, P. White, interviews with author; *Briefing: Falling through the Cracks*, 2005.

p.50 In 2005, Janice got a realtor's license. . . . J. Ramsey, interviews with author. Janice thought it would be difficult to sell three mortgages, since—because of state regulations—it was extremely difficult to sell a mortgage to somebody she was representing as a home buyer, too.

p.50 In the last few years, scientists have begun learning . . . Nuffield Council on Bioethics, *Pharmacogenetics: Ethical Issues* (London, September 2003), pp. 11–18.

p.51 . . . how a person's genetic idiosyncrasies affect the way that person reacts to different medications . . . Scientists have discovered, for example, that a tiny mutation in a gene that produces a liver enzyme called CYP2D6 can dramatically alter the effectiveness of certain drugs, such as the painkiller codeine. Some people with low levels of the enzyme won't process codeine at all; others will suffer serious side effects. By identifying those who have the mutation, doctors can avoid giving codeine to them—or at least can adjust for the side effects. See, for example, J. Kirchheiner, H. Schmidt, M. Tzvetkov, J-Tha Keulen, J. Lötsch, I. Roots and J. Brockmöller, "Pharmacokinetics of codeine and its metabolite morphine in ultra-rapid metabolizers due to *CYP2D6* duplication," *The Pharmacogenomics Journal*, July 1, 2006, pp. 1–9.

p.51 But as genetic testing improves . . . American Academy of Actuaries, "Genetic Information and Medical Expense Insurance," Public Policy Monograph (Washington, D.C.), June 2000; Paul L. Brockett, Richard MacMinn, and Maureen Carter, "Genetic Testing, Insurance Economics, and Societal Responsibility," *North American Actuarial Journal*, January 1999, Vol. 3, No. 1, pp. 1–20; Mark Hall, "Restricting Insurers' Use of Genetic Information: A Guide to Public Policy," *North American Actuarial Journal*, January 1999, pp. 34–46; Ellwood F. Oakley III, "Federal Regulation of Use of Genetic Information by Insurers: What Constitutes Unfair Discrimination?" *North American Actuarial Journal*, pp. 116–132; Phillip Longman and Shannon Brownlee, "The Genetic Surprise," *Wilson Quarterly*, October 1, 2000; Kathy L. Hudson, Karen H. Rothenberg, Lori B. Andrews, Mary Jo Ellis Kahn, Francis S. Collins, "Genetic Discrimination and Health Insurance: An Urgent Need for Reform," *Science*, Vol 27, October 20, 1995. In the last few years, the federal government and most of the states have passed laws regulating how and when insurance companies may consider genetic testing as part of the underwriting process. But the regulations vary widely from state to state and the federal requirements, which are part of HIPAA, apply only to group policies. For more on the present state of regulation, see also Alissa Johnson, "Genetics and Health Insurance," Briefing Papers on Important Genetic Issues of the Day (Washington: National Conference of State Legislatures, Genetic Technologies Project), August 2004.

p.51 "The insurance industry's great fear is that genetic testing . . . Charles S. Jones, Jr., "The Current State of Genetic Testing: An Insurance Industry Perspective on the Rush to Legislate," *North American Actuarial Journal*, Vol. 3., No. 1, p. 62. Jones was medical director for the Life Insurance Company of Georgia.

p.52 . . . insurers in the individual and small-business markets cross-reference applications . . . Bob Litell, "The Mysterious MIB—Not Any More . . ." *National Underwriter*, February 7, 2000; Douglas M. Mertz, "Avoiding Misrepresentation of Medical History," *AHIP* (Magazine of America's Health Insurance Plans), March/April 2006, pp. 57–58; Julie Miller, "Fraud Finding," *Managed Healthcare Executive*, September 2005; Merlis, pp. 7–8. Insurers ask that applicants waive their right to confidentiality in order to permit access to these records. Applicants are under no obligation to do so, but insurers generally won't offer coverage to anybody who refuses to go along.

p.52 . . . insurers' ability to actually use such information is still somewhat crude. . . . Ian Duncan and Ruth Ann Woodley, interviews with author. (Duncan and Woodley are actuaries based in Hartford, Connecticut.)

p.53 . . . some women with family histories of breast cancer . . . Emily A. Peterson, Kara J. Milliron, Karen E. Lewis, Susan D. Goold, and Sofia D. Merajver, "Health Insurance and Discrimination Concerns and BRCA1/2 Testing in a Clinic

Population," *Cancer Epidemiology, Biomarkers and Prevention*, Vol. 11, January 2002, pp. 29–87.

p.53 . . . giving insurers more leeway, not less . . . This is based primarily on my reporting about the "Health Care Choice Act." ("Pick and Lose," *TNR*, August 22, 2005.)

CHAPTER 3: AUSTIN

p.55 The city of Lakeway sits on the southern banks . . . Author's visit to Lakeway; A. Denys Cadman, and Byron D. Varner, *"In the Beginning": A Brief History of Lakeway* (Austin: Silverdale, 1981). Viewed via The Handbook of Texas Online at www.tsha.utexas.edu/handbook/online/articles/LL/hll80.html; Don Cooper, "Land Rush on Lakeway," *Austin American-Statesman*, May 4, 1987; Tara Tower, "Lakeway Deer Dilemma: Should They Stay or Should They Go?" *Austin American-Statesman*, April 1, 1997.

p.56 In 1991, Elizabeth Hilsabeck craved the Lakeway lifestyle . . . Elizabeth Hilsabeck and Steven Hilsabeck, interviews with author; author's visit to Lakeway, Texas; Elizabeth also supplied photos of the home and the twins' time in the hospital during their first year.

p.60 By 1993, the year after Parker and Sarah were born . . . See John K. Iglehart, "Physicians and the Growth of Managed Care," *NEJM*, Vol. 331, No. 17, October 27, 1994, pp. 1167–1171; and Alliance for Health Reform, *Sourcebook for Journalists* (Washington, D.C., 2003), p. 106. By the end of the decade, the proportions would reverse, with more than 90 percent of workers enrolled in some type of managed care.

p.60 During the 1990s, the cost of health care rose more slowly . . . Katherine R. Levit, Helen C. Lazenby, and Lekha Sivarajan, "Health Care Spending in 1994: Slowest in Decades," *HA*, Vol. 15, No. 2, Summer 1996 pp. 130–144.

p.61 . . . most people in HMOs ultimately reaped benefits . . . David Cutler, *Your Money or Your Life: Strong Medicine for America's Health System* (New York: Oxford University Press, 2004), pp. 86–99.

p.61 . . . it's easy to forget its idealistic lineage . . . Paul Starr, *The Social Transformation of American Medicine: The Rise of a Sovereign Profession and the Making of a Vast Industry* (New York: Basic Books, 1982), pp. 303–304.

p.62 At roughly the same time that Shadid's clinic was opening in Oklahoma . . . Starr, pp. 321–325; Merwyn R. Greenlick, *Annals of the American Academy of*

Political and Social Science, Vol. 399, January 1972, pp. 100–113. (Greenlick was director of the Health Services Research Center at Kaiser Foundation Hospitals in Portland, Oregon.) Author's disclosure: Some of Kaiser's fortune eventually endowed the Kaiser Family Foundation, which underwrote a fellowship for me while I was writing this book. But the foundation has no connection to the Kaiser Permanente health system.

p.63 **The American Medical Association (AMA) greeted this development** . . . Starr, pp. 326–327. The courts would eventually rule that some of the AMA's efforts to impede group practices—in particular, its attacks on the Group Health Association, the trade group that represented such practices—violated antitrust laws.

p.63 **A California native, Ellwood had come east to Minneapolis** . . . Paul Ellwood, interview with author; Joseph L. Falkson, *HMOs and the Politics of Health System Reform* (Chicago, Ill.: American Hospital Association, 1980), p. 14. Ellwood was hardly the first health care reformer to notice the virtues of group practices. The Committee on the Costs of Medical Care had endorsed group practice, as did a number of experts at the time. See, for example, Joseph Collins, "Group Practice in Medicine," *Harper's Monthly Magazine*, Vol. 157, June 1928, p. 172; and C. Rufus Rorem, "The Costs of Medical Care," *Accounting Review*, Vol. 7, No. 1, March 1932, p. 39.

p.64 **Ellwood thought there were better ways** . . . Ellwood, interview with author. For more on the early history of managed care, see William A. MacColl, M.D., *Group Practice and Prepayment of Medical Care* (Washington, D.C.: Public Affairs Press, 1966), cited in Falkson, pp. 16–17.

p.64 **Ellwood began talking up group practices** . . . Ellwood, interview with author.

p.65 **It was based on voluntary participation** . . . Employers, on the other hand, would receive orders. Under the eventual HMO legislation that passed, the government mandated that all large employers offer workers an HMO—as a choice—if they offered other sorts of health insurance. For the full story of the HMO Act of 1973, from its origins to its final provisions, see Falkson's exhaustive account.

p.65 **He finally started to make headway in the 1980s** . . . Ellwood, interviews with author; George Anders, *Health against Wealth: HMOs and the Breakdown of Medical Trust*, paperback ed. (New York: Houghton Mifflin, 1996), pp. 16–34; Mary J. Pitzer et al., "HMO: The Three Letters That Are Revolutionizing Health Care," *BW*, March 2, 1987, p. 94. Donald L. Barlett and James B. Steele, *Critical Condition: How Health Care in America Became Big Business and Bad Medicine* (New York: Doubleday, 2004), pp. 88–90. For more on the evolution of the HMO under Nixon and Reagan, see Starr, pp. 396–397; Maggie Mahar, *Money-Driven Medicine: The Real Reason Health Care Costs So Much* (New York: HarperCollins, 2006), pp.22–25.

p.66 By 1997, for-profit companies had two-thirds of the HMO business ...
KFF, *Trends and Indicators in the Changing Health Care Marketplace, 2002–Chartbook* (Washington, D.C., 2003), p. 57.

p.66 Early on, some of these for-profit insurers ... Robert Kuttner, "Must Good HMOs Go Bad (First of Two Parts)," *NEJM*, May 21, 1998, pp. 1558–1563; Kuttner, "The American Health Care System: Wall Street and Health Care," *NEJM*, Vol. 340, No. 8, February 25, 1999, p. 666. See also David U. Himmelstein, Steffie Woolhandler, Ida Hellander, and Sidney Wolfe, "Quality of Care in Investor-Owned versus Not-for-Profit HMOs," *JAMA*, Vol. 282, 1999, pp. 159–163. For a warning about the potential importance—or lack thereof—of a health plan's medical loss ratio, see James C. Robinson, "Use and Abuse of the Medical Loss Ratio to Measure Health Plan Performance," *HA*, Vol. 16, No. 4, July/August 1997, pp. 176–187.

p.67 If the ratio crept higher ... For an example of this dynamic in action, see Meera Somasundaram, "Higher Medical Costs Hurt Aetna," *LAT*, May 11, 2001.

p.67 But even more important than ownership status ... Jon Gabel, "Ten Ways HMOs Have Changed in the 1990s," *HA*, Vol. 16, No. 3, May/June 1997, pp. 135–136. The prototype IPA was actually formed not long after the first group practices, in the San Francisco Bay area, as an effort by doctors to compete with Kaiser but in an organization with looser control and oversight. (Source: Starr, p. 325.)

p.67 But the distinction was crucial. ... Kuttner, p. 1560; R. S. Thompson, "What Have HMOs Learned about Clinical Prevention Services? An Examination of the Experience at Group Health Cooperative of Puget Sound," *Milbank Quarterly*, Vol. 74, 1996, pp. 469–509, cited in Kuttner; Steffie Woolhandler and David U. Himmelstein, "Extreme Risk: The New Corporate Proposition for Physicians," *NEJM*, Vol. 333, December 21, 1995, pp. 1706–1708; M. R. Gold et al., "A National Survey of the Arrangements Managed-Care Plans Make with Physicians," *NEJM*, Vol. 333, December 21, 1995, pp. 1678–1683.

p.68 ... its new one was Leonard Abramson ... Leonard Abramson, *Healing Our Health Care System* (New York: Grove Weidenfeld, 1990).

p.68 ... it also became notorious for its adversarial relationship with physicians ... John Fairhall, "HMO Changes Medicine with Tough Bargaining," *Baltimore Sun*, October 6, 1994; John George, "Interview with a Pioneer in Managed Care," *Philadelphia Business Journal*, October 27, 1995.

p.68 These efforts helped U.S. Healthcare ... Steve Sakson, "Aetna Acquiring U.S. Healthcare," *Houston Chronicle*, April 2, 1996.

p.69 The Texas Medical Association is among the strongest ... Mike Menichini,

"HMO Fever," *Texas Business*, August 1995, pp. 79–88. Citing data from In-terstudy Publications, the article states that as of 1984, just over 2 percent of Texans were in HMOs, compared with just over 6 percent nationwide. In 1995, the respective percentages were about 10 percent for Texas and 17 percent for the nation. For more on the Texas Medical Association, see Wayne Guglielmo, "America's Best Medical Society?" *Medical Economics*, August 6, 2001, p. 77.

p.69 . . . the allure of more affordable health insurance had simply become too strong . . . Travis E. Poling, "Searching for a Cure-All," *San Antonio Business Journal*, November 8, 1991, p. 13. As an example, the article cited an HMO in San Antonio offering workers monthly premiums of about $240 for a family policy, $60 less than coverage cost under the traditional Blue Cross plan. This dif-ferential was consistent with national figures. In 2001, the average annual premi-um for family coverage in a private, employer-provided HMO was about $6,500, more than $1,000 less than the premium in a traditional fee-for-service plan. KFF, *Trends and Indicators*, 2002, p. 26.

p.69 First it was Southland convenience stores . . . Laura Castaneda, "Prognosis for the Future: Companies Prod Workers toward Managed Care," *Dallas Morn-ing News*, December 5, 1993.

p.69 . . . only one-fourth of all Texans in medium- to large-size companies . . . Sherry Jacobson, "More Firms Expected to Turn to HMOs: Costs Spark Concern as Reforms Bog Down," *Dallas Morning News*, September 12, 1994.

p.69 The Hilsabecks had entered the world of managed care . . . E. Hilsabeck and S. Hilsabeck, interviews with author.

p.70 That was a pretty big deal . . . E. Hilsabeck and S. Hilsabeck, interviews with author. See also Deanna Ricks, "Parents Locked in Battle with HMO's," *Oak Hill Gazette*, August 3, 1995. Both Prudential and Aetna, the company that eventually acquired PruCare from Prudential, declined requests to discuss the spe-cifics of the Hilsabeck case.

p.72 Within a few weeks a letter from Prudential arrived . . . Letter from Candice Wilson, PruCare customer service supervisor, to E. Hilsabeck, April 27, 1994; E. Hilsabeck, interviews with author.

p.72 In the fall of 1994, the company refused to pay . . . Letter from Wm. Josiah Taylor, MD, MPH, Prudential Medical Director, to Carol Faget, MD, Austin, Texas, September 22, 1994; E. Hilsabeck, interviews with author.

p.72. . . by having Elizabeth take a part-time job . . . The job was as a customer service representative for Allstate insurance (automobile and property, not health). E. Hilsabeck, S. Hilsabeck, interviews with author; Ricks, *Oak Hill Gazette*, p. 1.

p.72 Getting different health insurance was no simple matter . . . E. Hilsabeck and S. Hilsabeck, interviews with author; unpublished transcript of speech by Elizabeth Hilsabeck to Texas Physical Therapy Association, p. 5. This story comes purely from her recollection. However, such a policy would likely have been consistent with the minimal federal guidelines for HMOs at the time. Those guidelines, according to a letter sent from the U.S. Department of Health and Human Services, stated that insurers must cover up to two months of rehabilitative services—but only if those services "can be expected to result in the significant improvement of the member's condition within a period of two months" and not for "long-term rehabilitation" beyond that time. (Source: Jean D. LeMasurier, letter to Gary Young, June 1, 1995. LeMasurier was director of Policy and Program Improvement at the Office of Managed Care, within HHS. Young worked for the Texas Office of Public Insurance Council.)

p.73 Steven kept his job . . . E. Hilsabeck and S. Hilsabeck, interviews with author.

p.74 Garbled communications with insurance companies' representatives. . . . Ellyn E. Spragins, "Beware Your HMO," *Newsweek*, October 23, 1995, p. 54.

p.74 There was the story of the Adams family . . . Anders, *Health against Wealth* pp. 1–13.

p.74 And then there was the story of Bryan Jones . . . William Sherman, "What They Didn't Know about HMOs May Have KILLED THIS BABY," *New York Post*, September 18, 1995.

p.75 A few of these cases led to successful lawsuits . . . Anders, p. 11; Erik Eckholm, "$89 Million Awarded Family Who Sued H.M.O.," *NYT*, December 30, 1993; Erik Larson, "The Soul of an HMO," *Time*, January 22, 1996, p. 44 Thomas Maier, "Speedy Delivery," *Newsday*, June 27, 1995; Maier, "Mom Testifies on Baby's Death," *Newsday*, October 19, 1995.

p.75. . . with a public already wary of managed care . . . As late as 1997, 34 percent of respondents to a Kaiser Family Foundation/Harvard School of Public Health Survey said they thought managed care plans did a "good job" while just 21 percent said they thought such plans did a "bad job." Within two years, after many horror stories about HMOs, the numbers had reversed, with 39 percent saying "bad job" and 24 percent saying "good job." KFF, *Trends and Indicators*, 2002, p. 79.

p.75 Public frustration expressed itself most famously . . . Milt Freudenheim, "Survey Shows Dominance of Managed Care Plans in 1997," *NYT*, January 20, 1998. Steven Findlay, "Ailing HMOs: Eroding Confidence, Losses Weren't Part óf HMO's Plan," *USA Today*, December 30, 1997; Ellis Henican, "A Health Plan to Make You Sick," *Newsday*, February 4, 1998.

p.75 **Scholars could find no substantial evidence** . . . Cutler, p. 91; R. H. Miller and H. S. Luft, "Does Managed Care Lead to Better or Worse Quality of Care?," *HA*, Vol. 16, No. 5, pp. 7–25.

p.75 . . . **true to their original spirit, HMOs covered more basic preventive services** . . . KFF, *Trends and Indicators*, 2002, p. 36.

pp. 75–76 **The tale about Bryan Jones** . . . Sherman, "What They Don't Know. . . ."

p.76 **In other cases, the treatment an HMO refused** . . . Denise Grady, "Doubts Raised on a Cancer Procedure," *NYT*, April 16, 1999; Michael M. Mello and Troyen A. Brennan, "The Controversy over High-Dose Chemotherapy with Autologous Bone Marrow Transplant for Breast Cancer," *HA*, Vol. 20, No. 5, September/October 2001, pp. 101–117. More generally, see Peter D. Jacobson and Shannon Brownlee, "The Health Insurance Industry and the Media: Why the Insurers Aren't Always Wrong," *Houston Journal of Health Law and Policy*, Vol. 5, December 2005, p. 237.

p.77 **That had been one goal of Clinton's health care plan, too.** Although Clinton's plan envisioned most Americans moving into managed care, it would also have regulated such care closely, establishing standard benefit packages and providing a formal, binding appeals process for beneficiaries who felt they'd been wrongly denied coverage for treatments or access to specialists. It also would have left decisions about coverage of new treatments to a commission that included consumer representatives and health care professionals. In effect, Clinton's plan proposed to set strict rules for how managed care could reduce costs. Those rules might make coverage marginally more expensive, but, the thinking went, it was a price worth paying in order to spare those people in the greatest medical and financial need arbitrary or excessive penny-pinching. (Source: Ira Magaziner, interviews with author.)

p.77 . . . **the guidelines of a consulting firm called Milliman and Roberts** . . . A series of studies suggested that Milliman's guidelines were not only out of step with standard practice but dangerously so; the company, while denying such charges, noted that the guidelines were meant merely as a tool for insurers to use in determining their own clinical standards. (Source: Jim Loughman, interview with author. Loughman was a spokesperson for Milliman.) See Jeffrey S. Harman and Kelly J. Kelleher, "Pediatric Length of Stay Guidelines and Routine Practice: The Case of Milliman and Robertson," *Archives of Pediatric and Adolescent Medicine*, Vol. 155, August 2001, pp. 885–890; Narenda M. Kini et al., "Inpatient Care for Uncomplicated Bronchitis: Comparison with Milliman and Robertson Guidelines," *Archives of Pediatric and Adolescent Medicine*, Vol. 155, December 2001, pp. 1323–1327; Marion R. Sills, "Pediatric Milliman and Robertson Length-of-Stay Criteria: Are They Realistic?" *Pediatrics*, Vol. 105, No. 4, April 2000, pp. 733–737. For more background on Milliman and the debate over its work, see George Anders and Laurie McGinley, "Medical Cop: Actuarial Firm Decides Just How Long You Spend in Hospital," *WSJ*, June 15, 1998; and Jonathan Gardner, "Dueling Studies: Lengths of Stay for Mastectomy Patients at Issue," *Modern Healthcare*, May 26, 1977.

p.77 Plans offered financial bonuses to doctors ... George D. Lundberg with James Stacey, *Severed Trust: Why American Medicine Hasn't Been Fixed* (New York: Basic Books, 2000), pp. 98–99.

p.78 One study in the *Archives of Internal Medicine* found ... Ron Winslow, "Hospitals Sending Hip Fractures Patients Home Too Soon," *Pittsburgh Post-Gazette*, January 21, 2003. A problem with virtually every study examining the outcomes in managed care and traditional medicine was the difficulty of accounting for the fact that different people tended to sign up for different kinds of plans in the first place, with more healthy people seeking managed care (because they shopped largely on the basis of price) and with more unhealthy people seeking traditional coverage (because they shopped largely on the basis of unfettered access to doctors and hospitals). But one crucial study did manage to adjust for the health of its patients. The research, published in the *Journal of the American Medical Association*, looked at where patients in New York got a certain type of cardiac procedure. It found that during the 1990s, managed care was far more likely to send its patients to hospitals with relatively high mortality rates—just because these happened to be the hospitals with the cheapest prices. See Lars C. Erikson et al., "The Relationship between Managed Care Insurance and Use of Lower-Mortality Hospitals for CABG Surgery," *JAMA*, Vol. 283, No. 15, April 19, 2000, pp. 1976–1982.

p.78 Plans like Group Health in Seattle had embraced ... Charles Ornstein and Tracy Weber, "Kaiser Put Kidney Patients at Risk," *LAT*, May 3, 2006; "Kaiser Denied Transplants of Ideally Matched Kidneys," May 4, 2006, and "Kaiser Official Apologizes," May 11, 2006; Kuttner, "Must Good HMOs Go Bad?" p. 1560.

p.79 ... particularly distressing to Paul Ellwood himself ... Christine Gorman, "Playing the HMO Game," *Time*, July 13, 1998; Ellwood, interviews with author.

p.79 ... Ellwood also had a personal perspective ... Lisa Belkin, "The Ellwoods: But What about Quality?" *New York Times Magazine*, December 8, 1996, p. 68; Ellwood, interviews with author.

pp. 79-80 ... joined in a class action lawsuit against the nation's top insurers ...

Three of the defendants—Aetna, Cigna, and WellPoint—eventually settled the lawsuits, paying out hundreds of millions of dollars to the plantiffs. J. C. Conklin, "Doctors' Fee Suit Criticizes Cigna," *Dallas Morning News*, October 2, 2001; Milt Freudenheim, "WellPoint Agrees to Settle a Long Dispute with Doctors," *NYT*, July 12, 2005; Denise Gellene, "Doctors Press Battle against Insurers," *LAT*, March 27, 2001; Joseph B. Treaster, "Aetna to Settle Suit with Doctors over Payments," *NYT*, May 22, 2003.

p.80 . . . Elizabeth Hilsabeck felt that this was exactly what had been happening . . . E. Hilsabeck and S. Hilsabeck, interviews with author.

p.80 They complained to the bank's benefit manager . . . Letter from Doris Morriss, Human Resources Services Manager, Farm Credit Bank of Texas, to Susan Short, Account Representative, PruCare, April 26, 1995.

p.81 PruCare eventually backed down on this requirement . . . Laura Johannes, "State Probes HMO's Denial of Some Care," *WSJ* (Texas edition), August 16, 1995; "PruCare Reverses Policies toward Disabled Children," *Insurance News* (Texas edition), October 1995, p. 19.

p.81 . . . the entire managed care system operated in a sort of legal no-man's-land . . . This is based primarily on my reporting about managed care reform in the late 1990s. ("Mangling Care," *TNR*, August 10, 1998, p. 16; and "Cosmetic Surgery," *TNR*, August 17, 1988, p. 25.) See also Robert Pear, "HMO's Using Federal Law to Deflect Malpractice Suits," *NYT*, November 17, 1996.

p.81 Eventually Elizabeth got the idea to form an advocacy group . . . Connie Barron, Kathy Uhouse, E. Hilsabeck, and S. Hilsabeck, interviews with author; promotional material, literature, notes on telephone conversations, and correspondence relating to the Texas Advocates for Special Needs Kids, provided by E. Hilsabeck. (Barron was an official with the Texas Medical Association; Uhouse was a charter member of TASK.)

p.82 . . . the legislature finally passed a measure to reform managed care . . . This is based heavily on my reporting during the presidential campaign of 2000. ("Yuck, Yuck," *TNR*, November 6, 2000, p. 23.) See also Richard A. Oppel and Jim Yardley, "Bush Calls Himself Reformer; the Record Shows Label May Be a Stretch," *NYT*, March 20, 2000.

p.83 . . . Elizabeth, a lifelong Republican, voted for Bush . . . E. Hilsabeck, interview with author.

p.83 But once he got to the White House . . . Amy Goldstein, "For Patients' Rights, a Quiet Fadeway," *WP*, September 12, 2003.

p.83 Back in Texas, some health professionals said they noticed . . . Tom Banning, interview with author. (Banning was an official with the Texas Academy of Family Physicians.) See also Carol Marie Cropper, "In Texas, a Laboratory Test on the Effects of Suing HMOs," *NYT*, September 13, 1998; Dawn Mackeen, "Woe is HMO," *Salon*, October 14, 1999.

p.83 In 2004, the U.S. Supreme Court invalidated . . . Linda Greenhouse, "Justices Limit Ability to Sue Health Plans," *NYT*, June 22, 2004.

p.84 Managed care by then had already adapted on its own . . . Allison Bell, "Ad Campaign Boosts Managed Care's Image," *National Underwriter*, September 28, 1998, p. 26; Gabel, "Ten Ways HMOs Have Changed," pp. 134–145; James C. Robinson, "From Managed Care to Consumer Health Insurance: The Fall and Rise of Aetna," *HA*, Vol. 23, No. 2, pp. 43–55; Barbara Martinez, "Tired of Being Cast as the Villain, HMOs Hire Talent Agency," *WSJ*, July 9, 2002.

p.84 As for Elizabeth . . . E. Hilsabeck, Parker Hilsabeck, and S. Hilsabeck, interviews with author.

CHAPTER 4: SIOUX FALLS

p.87 Morning shift inside the pork by-products room . . . Ben Aning, Clinton Baldwin, and Lester Sampson, interviews with author. (Aning and Baldwin were retirees from Morrell.)

p.88 Lester, still just seventeen years old, was technically . . . Audrey Sampson and L. Sampson, interviews with author.

p.89 . . . that was a deal workers like Lester gladly made . . . B. Aning, A. Sampson, and L. Sampson, interviews with author.

p.90 But the security Morrell provided turned out to be fleeting . . . B. Aning, C. Baldwin, Jim Larson, A. Sampson, L. Sampson, and Gerrit Zwak, interviews with author (Zwak was a retiree from Morrell, and Larson was a union official); Gigi Verna, "Morrell Cuts Benefits to 2,800 Ex-Employees," *Greater Cincinnati Business Record*, February 6, 1995; Linda Eardley, "Retirees Stripped of Health Coverage," *St. Louis Post-Dispatch*, February 17, 1995.

p.90 Morrell offered a straightforward rationale . . . "For Iowa Retirees, a Promise Goes on the Chopping Block," *WP*, March 5, 1995.

p.90 . . . it had become customary for large companies . . . *Current Trends and Future Outlook for Retiree Health Benefits: Findings from the Kaiser/Hewitt 2004 Survey on Retiree Health Benefits* (Washington, D.C.: KFF, December 2004), p. v; Patricia Neuman et al., *Medicare Chartbook* (Washington, D.C.:

KFF, 2005), p. 23; Jonathan Gelb, "Scrambling for Health Benefits," *Philadelphia Inquirer*, February 13, 2003; Brett Clanton, "Big Three Scale Back Benefits," *Detroit News*, April 10, 2005; Danny Hakim, "GM and Union in a Deal to Cut Health Benefits," *NYT*, October 18, 2005.

p.90 But Medicare, a program frequently criticized for its supposedly excessive generosity . . . *Fact Sheet: Medicare at a Glance* (Washington, D.C.: KFF, September 2005). Technically, the cost sharing in Medicare is different from the co-payments that private insurance plans frequently feature. Cost sharing means that beneficiaries pay a percentage of charges, whereas co-payments mean that beneficiaries pay a fixed fee for each medical service rendered. It's an important distinction for the chronically ill, who can pay substantially more through cost sharing because of their high medical bills. (Source: Jonathan Oberlander, correspondence with author. Oberlander is a professor of political science at the University of North Carolina and the author of *The Political Life of Medicare*.)

p.91 Since the 1970s, these out-of-pocket costs have risen . . . According to figures from the House Ways and Means Committee, from 1972 to 1988, out-of-pocket medical spending for seniors rose from 7.8 percent of after-tax household income to 12.5 percent. Another set of calculations, by Medicare expert Marilyn Moon, produced somewhat different figures: According to Moon, out-of-pocket spending as a percentage of income among elderly Medicare beneficiaries fell from 19.1 percent in 1965 to 11 percent in 1970, then climbed steadily until 2004, the last year for which she calculated, when it hit 22.6 percent. While her figures are different—perhaps because Ways and Means used after-tax income—the pattern is the same: after plummeting with the start of Medicare in 1966, out-of-pocket expenditures rose steadily over time. Indeed, if Moon's figures are correct, then by 2004 seniors were actually paying more of their health care expenses directly than they were in the days before Medicare existed. (Sources: John Iglehart, "The American Health Care System: Medicare," *NEJM*, Vol. 327, No. 20, November 12, 1992, p. 1469; Marilyn Moon, *Medicare: A Policy Primer* (Washington: Urban Institute Press), 2006, p. 23.

p.91 More affluent seniors have been able . . . Iglehart, p. 1469; *Medicare Chartbook*, pp. 20–24, 32–37.

p.91 . . . forced the "special problem of the elderly" onto center stage . . . Herman Miles Somers and Anne Ramsey Somers, *Doctors, Patients, and Health Insurance: The Organization and Financing of Medical Care* (Washington, D.C.: Brookings Institution, 1961), 426.

p.91 But the elderly were not sharing in this progress . . . Somers and Somers, pp. 407–408, 426–444; Jill Quadagno, *One Nation Uninsured: Why the U.S. Has No National Health Insurance* (New York: Oxford University Press, 2005) pp. 61–62. Note that the Blue Cross system behaved in a conspicuously different way. Most of the Blues did allow policyholders to extend their insurance into their retirement years. But the Blues were already paying a heavy price for such liberal enrollment policies. The willingness to offer, or at least maintain, coverage for the elderly was a major factor in the adverse selection "death spiral" that—as discussed in Chapter 3—eventually prompted the Blues to abandon community rating and other features designed to make its insurance available to everybody.

p.92 In Washington, the frustrated proponents . . . Theodore R. Marmor, *The Politics of Medicare*, 2nd ed. (New York: Aldine de Gruyter, 2000), pp. 9–11.

p.92 In the 1960s, political momentum finally seemed to swing . . . Marmor, pp. 31–56; "Senate Kills Medicare, 52–48; Kennedy Calls for a Congress in '63 That Will Enact Plan," *WP*, July 18, 1962.

p.92 But the architects of the Medicare proposal . . . Marmor, pp. 10–21; Jonathan Oberlander, *The Political Life of Medicare*, (Chicago, Ill.: University of Chicago Press, 2003), pp. 22–40.

p.93 Although the doctors never backed down . . . Quadagno, p. 72.

p.93 Soon there was a huge increase in the use of medical services . . . Arthur Hess, "Medicare after One Year," *Journal of Risk and Insurance*, Vol. 35, No. 1, March 1968, p. 124. (Hess was deputy commissioner of the Social Security Administration.)

p.93 Eventually the cost of Medicare became so big . . . Oberlander, pp. 120–133; "Panel Urges More Uniform Medicare Fees for Physicians," *Philadelphia Inquirer*, March 5, 1987.

p.94 These reforms succeeded in their primary goal . . . *Medicare Chartbook*, 2005, 58. The effects of these and subsequent payment reforms on the actual quality of health care remain controversial—with anecdotes about seniors leaving the hospital "quicker and sicker" on the one hand and studies suggesting no serious health effects on the other. (See Charles Green, "Panel: Elderly Leaving Hospitals 'Quicker and Sicker,'" *Philadelphia Inquirer*, November 30, 1989.)

p.94 . . . the standard private insurance policies for working-age Americans expanded . . . Ernst R. Berndt, "Why Major Growth in Times of Cost Containment?," *HA*, July/August 1999; Oberlander, pp. 36–55. Outpatient care became particularly important as hospitals, under pressure to cut costs, began discharging people earlier.

p.94 . . . establishing Medicare's first "catastrophic benefit" package . . . Marmor, pp. 110–113; Oberlander, pp. 56–73; Quadagno, pp. 149–158; Bill Peterson, "Rostenkowski Heckled by Senior Citizens," *WP*, August 18, 1989.

p.95 When Lester Sampson's father was working at Morrell ... A. Sampson and L. Sampson, interviews with author.

p.96 ... a local from the Amalgamated Meat Cutters and Butcher Workers of North America ... Walter A. Simmons, "Brute Cries of Rage Fill Air as Men Meet," *Argus Leader*, 1935. (Exact date unknown.)

p.96 In the eventual settlement, Morrell's workers won ... "Unions and Morrell Settle Disputes," *Argus Leader*, March, 1937; "Poor Pay, Conditions Forced the Big Strike," *Argus Leader*, 1978.

p.96 To hear Lester and some of his colleagues tell it ... B. Aning, C. Baldwin, Bruce Iwerks, George Iwerks, Frances Krier, Jim Larson, Sally Linton, A. Sampson, L. Sampson, and G. Zwak, interviews with author. See also Steve Young, "Retired Morrell Workers Enjoy Slice of Old Times," *Argus Leader*, November 27, 1989; and Mark Trautmann, "City Asks Morrell to Curtail Stench," *Argus Leader*, May 7, 1991.

p.97 But the business of meatpacking turned out to be no different ... Tony Brown, "Changes on the Horizon for Morrell," *Argus Leader*, September 6, 1981; Mel Antonen, "Pork Packers: It's Survival of Fittest," *Argus Leader*, August 5, 1983.

p.98 Morrell had quietly undergone a second ... Clare M. Reckert, "1970 Loss Posted by United Brands," *NYT*, April 7, 1971.

p.98 At first, the union in Sioux Falls grudgingly went along .... Antonen, "Workers Avoid Layoffs with Pay Cuts," *Argus Leader*, September 26, 1983; Frederic J. Hron, "Another Pay Cut Looms at Morrell," *Argus Leader*, September 10, 1984; Hron, "A Year Later, Morrell Workers Making Ends Meet—Barely," *Argus Leader*, October 14, 1984; Anne Willette, "Morrell Rejects Wage Plan," *Argus Leader*, August 28, 1985; Steve Erpenbach, "Morrell Says It Seeks Competitive Wage," *Argus Leader*, September 14, 1988; Brenda Wade Schmidt, "Morrell Workers Sue Their Union," *Argus Leader*, January 23, 1992; "Aftermath of Walkout Stretches into 1992," *Argus Leader*, May 4, 1992.

p.98 The one place where the union had held the line ... "Morrell Contract Has Less Pay, More Benefits," *Argus Leader*, September 26, 1983; Schmidt, "Union OKs Morrell Pact 2-to-1," *Argus Leader*, January 29, 1991; "Aftermath," *Argus Leader*, 1992; Trautmann, "Morrell Files Suit to Cut Retiree Benefits," *Argus Leader*, December 19, 1991.

p.99 ... the price of all benefits, for present and retired workers, nearly doubled Morrell's labor costs ... "Closings Will Cost S.F. Workers' Jobs," *Argus Leader*, September 6, 1981.

p.99 ... a change in the federal rules governing company benefits ... P. Neuman, correspondence with author; Larry Light, "Honest Balance Sheets, Broken Promises," *Business Week*, November 23, 1992, p. 106.

p.99 Morrell was one of those. . . . B. Aning, C. Baldwin, L. Sampson, and G. Zwak, interviews with author; Trautmann, "Morrell Files Suit," *Argus Leader*, 1991. *John Morrell & Co. v. United Food and Commercial Workers International Union, AFL-CIO; Bernard J. Aning, as representative of a defendant class*, 37 F.3d 1302, filed October 12, 1994.

p.100 Armed with that decision . . . Trautmann, "Judge Sides with Morrell on Benefits," *Argus Leader*, June 26, 1993; Schmidt, "Morrell Eliminates Retirees' Insurance," *Argus Leader*, January 25, 1995; Schmidt, "High Court Won't Hear Retiree Plea," *Argus Leader*, May 31, 1995.

p.100 Once again the union officials scrambled . . . B. Aning, G. Larson, A. Sampson, L. Sampson, and G. Zwak, interviews with author.

p.100 Clinton Baldwin had worked at Morrell . . . B. Aning and C. Baldwin, interviews with author.

p.102 James Couch had a harder time . . . James Couch Jr. and Emit Van Veen, interviews with author.

p.103 The Sampsons were far more typical. . . . A. Sampson and L. Sampson, interviews with author. The Sampsons could have purchased Blue Cross insurance that would have covered their prescriptions. However, according to Audrey, the coverage would have been partial—and the out-of-pocket payments, combined with the higher premiums, would have left them no better off financially.

p.104 For a while, the annual trips . . . A. Sampson and L. Sampson, interviews with author. See also Duncan Campbell, "This Is Nogales, Mexico," *Guardian*, May 2, 2002. In 2003, according to one study, 5 percent of all seniors purchased drugs from either Canada or Mexico. (See John Strahinich, "Study: Up to 40% Skip Meds," *Boston Herald*, April 19, 2005.)

p.105 But their truck wore down from all the traveling. . . . A. Sampson and L. Sampson, interviews with author.

p.106 Morrell's decision to cancel health insurance for retirees drew national attention. . . . Ellen E. Schultz, "Companies Quietly Use Mergers, Spinoffs to Cut Worker Benefits," *WSJ*, December 27, 2000; U.S. Senate Committee on Labor and Human Resources, *Health Insurance Coverage for 55- to 64-Year-Olds: Hearing*, 105th Cong., 2d sess., June 25, 1998.

p.106 . . . the number of employers offering retirement insurance . . . Paul Fronstin, *The Impact of the Erosion of Retiree Health Benefits on Workers and Retir-*

276

ees, EBRI Issue Brief 279 (Washington, D.C.: EBRI, March 2005); GAO, *Retiree Health Insurance: Erosion in Employer-Based Health Benefits for Early Retirees* (Washington, D.C.: July 11, 1997).

p.106 And in a few high-profile cases . . . *Retired Steelworkers and the Health Benefits: Results of a 2004 Survey* (Menlo Park, Calif.: KFF, May 2006), p. 6.

p.107 Polls found that the public generally supported this idea. . . . "The Public on Medicare Part D—the Medicare Prescription Drug Benefit," Kaiser Public Opinion Spotlight (Washington, D.C.: KFF, April 2006), pp. 3–4. See also Pew Center for Research, "Less Opposition to Gay Marriage, Adoption, and Military Service," March 22, 2006.

p.107 . . . the case for helping with drugs raised eyebrows among some pundits and politicians . . . Robert J. Samuelson, "A Costly Freebie," *WP,* June 4, 2003.

p.108 . . . the very same argument made by critics of Medicare in the early 1960s . . . Marmor, pp. 26–30.

p.108 . . . renewed faith in the private sector had displaced faith in government . . . Robert J. Blendon, Mollyann Brodie, and John Benson, "What Happened to Americans' Support for the Clinton Health Plan," *HA,* Summer 1995, pp. 12–13.

p.108 . . . Clinton and the Republican Congress agreed to create "Medicare + Choice." . . . Carlos Zarabozo, "Milestones in Medicare Managed Care," *Health Care Financing Review,* Vol. 22, No. 1, Fall 2000, pp. 65–66; KFF, "Medicare Fact Sheet: Medicare + Choice," (Washington, D.C.: KFF, April 2003). This also seemed to fulfill the vision of Paul Ellwood, architect of the act of 1973, who'd always thought that the way to spread managed care was to get Medicare heavily involved with it. Ellwood, interviews with author.

p.109 . . . the verdict on their finances was quite clear . . . GAO, "Medicare + Choice: Payments Exceed Cost of Fee-for-Service Benefits, Adding Billions to Spending," (Washington, D.C.: August 11, 2000), pp. 3–14.

p.109 . . . the HMOs started pulling back . . . Although the insurance industry would cite the reductions in payments as the primary reason for the withdrawals, other experts suggested that they also reflected the normal fluctuations of the insurance business cycle, in which insurers pull back on coverage and raise premiums once they've accumulated a large group of beneficiaries, in order to winnow out less profitable lines of business. Randall S. Brown and Marsha R. Gold, "What Drives Medicare Managed Care Growth?," *HA,* Vol. 18, no. 6, November/December 1999, pp. 143–148; Nora Super Jones, "Medicare HMO Pullouts: What Do They Portend for the Future of Medicare + Choice?," National Health Policy Forum, Issue Brief No. 730, February 5, 1999.

p. 109 Some cut back on prescription drug coverage . . . Lori Achman and Gold, *Medicare + Choice Plans Continue to Shift More Costs to Enrollees*, report prepared by Mathematica Policy Research, Inc., for Commonwealth Fund (New York, April 2003). Cited in Thomas R. Oliver, Philip R. Lee, and Helene L. Lipton, "A Political History of Medicare and Prescription Drug Coverage," *Milbank Quarterly*, Vol. 82, No. 2, 2004, p. 305.

p. 109 Others simply dropped out. . . . KFF, "The Role of National Firms in Medicare + Choice," June 2002, pp. 3–4.

pp. 109–110 . . . Bush signed into law the Medicare Modernization Act . . . George W. Bush, president's radio address, White House Radio, November 12, 2005; KFF, "Medicare Fact Sheet: Medicare Prescription Drug Plan, by State, 2006," November 2005; Health Policy Alternatives, Inc. for the Henry J. Kaiser Family Foundation, "Prescription Drug Coverage for Medicare Beneficiaries: Summary of the Final Rule to Implement the Prescription Drug Benefit (Title I of the Medicare Modernization Act of 2003)," February 10, 2005; Robin Toner, "Rival Visions Led to Rocky Start for Drug Benefit," *NYT*, February 6, 2006; "Key Provisions of the New Medicare Law," *Christian Science Monitor*, December 4, 2003.

p. 110 On a blustery December day in 2005 . . . A. Sampson, L. Sampson, interviews with author; "Overview of Tracleer Treatment," Tracleer Info Center, at http://www.tracleer-pph.com/pages/tracleer_info.html; and RxUSA listing for Tracleer, at http://rxusa.com/. The price for sixty 125-milligram tablets of Tracleer was about $3,700, or about $62 per tablet.

p. 111 But figuring this out was no easy task. . . . The law did require plans to carry at least two drugs in every therapeutic category, but there was always the danger that, as with Tracleer, electing a specific plan would mean switching to prescriptions that didn't work as well.

p. 111 . . . experts had warned that seniors would have trouble . . . Jeffrey S. Crowley, Testimony before the Senate Special Committee on Aging, June 28, 2005; Richard Jensen, "The New Medicare Prescription Drug Law: Issues for Enrolling Dual Eligibles into Drug Plans," Kaiser Commission on Medicaid and the Uninsured Issue Paper (Washington, D.C.: KFF, January 2005).

p. 111 But then the program actually began . . . Ceci Connolly, "The States Step In as Medicare Falters," *WP*, January 14, 2004; Robert Pear, "States Intervene after Drug Plan Hits Early Snags," *NYT*, January 8, 2006; Nancy Weaver Teichert, "Rx Emergency Declared," *Sacramento Bee*, January 13, 2006.

p. 112 The contrast to the implementation of the original Medicare program . . . Robert Ball and T. Marmor, interviews with author (Ball had served as commissioner of Social Security under Johnson; Marmor worked for the Department of Health, Education, and Welfare before going on to write *The Politics of Medicare*, widely considered the program's definitive history). Dan Morgan and Martin Weil, "Medicare Takes Over Easily," *WP*, July 2, 1966; Harold Schmeck, "Medicare's Start Has Been Smooth," *NYT*, July 25, 1966.

p.112. . . months before anybody could plausibly use the word "smooth" . . . Julie Rovner, "Up and Down," *National Journal's Congress Daily*, May 5, 2006.

p.113 . . . less about its workability than its sustainability . . . This is based primarily on my reporting of the debate over the Medicare drug benefit. ("Careless: The Medicare Law's True Cost," *TNR*, December 15, 2003.) The estimates of extra cost in the Bush Medicare drug program come from economist Dean Baker. See Baker, "The Savings from an Efficient Medicare Drug Plan," Center for Economic and Policy Research, Washington, D.C., January 2006. (Baker is co-director of the Center.)

CHAPTER 5: LAWRENCE COUNTY

p.115 Wanda Maldonado watched as Ernie tossed and turned . . . Wanda Maldonado, Donna Jones, Pat Burks, and Wanda Andrews, interviews with author; medical files of Ernesto Maldonado at the North Terrace Family Clinic in Lawrenceburg, Tennessee; John Spragens, "The Faces of TennCare," *Nashville Scene*, November 24, 2005.

p.117 Since its inception in 1965, Medicaid had . . . John K. Iglehart, "Medicaid," *NEJM*, March 25, 1993.

p.117 Tennessee's version of the program, called TennCare . . . Rick Lyman, "Once a Model, Tennessee's Health Plan Is Endangered," *NYT*, November 20, 2004.

p.118 . . . leaving the couple to beg for medicinal handouts . . . W. Maldonado, D. Jones, P. Burks, and W. Andrews, interviews with author.

p.118 . . . helping the poor was barely an afterthought . . . Milton Terris, "Medical Care for the Needy and the Medically Needy," *Annals of the American Academy of Political and Social Science*, Vol. 273, *Medical Care for Americans*, January 1951, pp. 84–92. See also Robert Stevens and Rosemary Stevens, *Welfare Medicine in America: A Case Study of Medicaid*, 2nd ed. (New Brunswick, N.J.: Transaction, 2003), pp. 5–23.

p.119 . . . produced serious talk of having Washington do something to help low-income Americans get health care . . . Jonathan Oberlander, *The Political Life of Medicare* (Chicago, Ill.: University of Chicago Press, 2003), pp. 27–39.

p.120 That was consistent with another long-standing American belief . . . Albert Deutsch, "The Sick Poor in Colonial Times," *American Historical Review*, Vol. 46, No. 3, April 1941, pp. 560–562.

p.120 Backers of Kerr-Mills had predicted . . . Stevens and Stevens, p. 29.

p.120 It did nothing of the sort. . . . Jill Quadagno, *One Nation Uninsured: Why the U.S. Has No National Health Insurance* (New York: Oxford University Press, 2005), p. 60. As Quadagno notes, by 1963 just over half of the states had even started their own programs, and of those, just four were offering all possible benefits.

p.120 The AMA and its allies made a final stand . . . Oberlander, pp. 30–31.

p.120 It filled in the gaps of Medicare . . . Iglehart, "Medicaid."; Barbara Lyons and Samantha Artiga, correspondence with author. (Lyons and Artiga, both affiliated with the Kaiser Family Foundation, are policy experts on Medicaid.)

p.121 It was a remarkably ambitious reach . . . Colleen Grogan and Eric Patashnik, "Between Welfare Medicine and Mainstream Entitlement: Medicaid at the Political Crossroads," *Journal of Health Policy, Politics, and the Law*, Vol. 28, No. 5, October 2003, pp. 825–828. As it turned out, one of the most important forces behind Medicaid's skyrocketing costs in the 1970s and 1980s was nursing home care, which Medicare did not cover and, because of its enormous expense, was quickly overwhelming the savings of even relatively affluent seniors. Ellen O'Brien, "Long-Term Care: Understanding Medicaid's Role for the Elderly and Disabled," KFF, November 2005.

p.122 . . . officials like Congressman Henry Waxman kept pushing to expand the program . . . Grogan and Patashnik, 829–833; Dan Morgan, "Medicaid Costs Balloon into Fiscal 'Time Bomb,'" *WP*, January 30, 1994.

p.122 . . . an understanding that people with low incomes had special medical needs . . . Iglehart, "Medicaid."; Lyons, interview with author. Another example of an effective intervention was the addition of universal lead screening to the Medicaid package in the early 1990s. Low-income children are at much greater risk of lead poisoning, which can lead to permanent brain damage, because of old, lead-based paint in dilapidated housing. Source: GAO, *Medicaid: Elevated Blood Levels in Children* (Washington, D.C., February 1998), p. 3.

p.123 Reagan wanted to make it a "block grant" . . . Helen Dewar, "Dominated by Reagan, Session Makes Much History in a Hurry," *WP*, December 17, 1981; B. Drummond Ayres Jr., "Reagan and Cuts in Revenue Sharing," *NYT*, October 2, 1981.

p.124 As a young man in Puerto Rico . . . W. Maldonado and D. Jones, interviews with author.

p.126 Lawrenceburg, a town of a just over 10,000 people . . . City of Lawrence-burg, official website, at http://www.cityoflawrenceburgtn.com/Demographics. htm.

p.126 But Wanda's enthusiasm for coming home was dampened . . . W. Maldonado and D. Jones, interviews with author; E. Maldonado's medical files.

p.127 In 1992, Ned McWherter was in his fifth year as governor . . . Ana Byrd Davis, "Runaway Costs, Wary Taxpayers Have State in Trap," (Memphis) *Commercial Appeal*, February 14, 1993; George Anders, *Health against Wealth: HMOs and the Breakdown in Medical Trust* paperback ed., (New York: Houghton Mifflin, 1996), pp. 190–209.

p.128 . . . many were simply disconnected from the health care system . . . Lyons, interview with author.

p.128 Aware that the public might be wary . . . David Brown, "Deluged by Medicaid, States Open Wider Umbrellas," *WP*, June 9, 1996; (Memphis) *Commercial Appeal*, "TennCare Answers," January 1, 1994.

p.129 Ernie and Wanda were among those who . . . W. Maldonado and P. Burks, interviews with author.

p.130 "Tenncare was great." W. Maldonado, interview with author.

p.130 The early months of the program had been chaotic . . . David Keim, "1 Change Not Enough for Some Doctors, Network Rule on TennCare Ripped," *Knoxville News-Sentinel*, January 13, 1994. In this respect, the experience proved to be an eerie precursor of what would happen with President Bush's Medicare drug plan, which—not coincidentally—also sought to use private insurance as a way to deliver benefits traditionally handled by the government itself.

p.130 . . . some had applied a simpler strategy . . . Martin Gottlieb, "A Free-for-All in Swapping Medicaid for Managed Care," *NYT*, October 2, 1995. Meanwhile, the HMOs affiliated with university hospitals were attracting much sicker patients, who wanted access to high-tech treatments, and so these HMOs were doing worse. (Source: Anders, *Health against Wealth*, p. 198.)

p.131 In one particularly egregious episode . . . Anders, pp. 204–205.

p.131 Many states skimped on administration and oversight . . . MEDPAC, "Report to the Congress: Select Medicare Issues," June 1999, pp. 19–33.

p.131 . . . reduced the payments they made for individual medical services . . .

Lyons, interview with author; MEDPAC, "2002 Survey of Physicians about the Medicare Program," March 3, 2002; MEDPAC, "Report to the Congress: Medicare Payment Policy," March 2005.

p.132 The same basic story line was unfolding across the country . . . "Managing Care for the Poor," WP, February 17, 1999; "Vermont Deal Seeks to Curb Medicaid Costs," Associated Press, October 3, 2005.

p.132 Some of the most egregious stories came from Florida . . . Fred Schulte and Jenni Bergal, "Profits from Pain," Sun-Sentinel, December 11, 1994; Michael Samai, "The Human Cost," (Fort Lauderdale, Fl.) Sun-Sentinel, December 11, 1994.

p.133 That initiative was called the State Children's Health Insurance Program . . . John M. Broder, "Health Coverage of Young Widens with States' Aid," NYT, December 4, 2005; Jennifer Preston, "New Jersey Republicans Choose Charity-Care Plan," NYT, May 2, 1996; Dan Bustard, "Vermont Ranks High on Health Care Efforts," (Claremont, N.H.) Eagle Times, November 15, 2004; Glenn Howatt, "State Approves Plan for Tobacco Money," (Minneapolis) Star Tribune, June 6, 2002.

p.133 . . . by this time, they had money to spare . . . Richard Wolf, "States Find Many Ways to Spend Surpluses," USA Today, May 27, 1998.

p.133 Even better, lawsuits against the tobacco industry . . . Barry Meier, "Tobacco Windfall Begins Tug-of-War among Lawmakers," NYT, January 10, 1999.

p.133 A few states with a strong progressive tradition . . . Dan Bustard, "Vermont Ranks High on Health Care Efforts," (Claremont, N.H.) Eagle Times, November 15, 2004; John Hoogesteger, "Tobacco Kitty Touted to Improve Health Care," St. Cloud (Minn.) Times, January 21, 1999; Glenn Howatt, "State Approves Plan for Tobacco Money," (Minneapolis) Star Tribune, June 6, 2002.

p.133 Among the states that would expand health insurance programs most aggressively . . . State of the States (Washington, D.C.: Academy for Health Research and Health Policy, January 2001), p. 4; Susan A. Riedinger, Amy-Ellen Duke, Robin E. Smith, and Karen C. Tumlin, Income Support and Social Services for Low-Income People in New Jersey (Washington, D.C.: Urban Institute, September 1996) p. 44. KidCare, New Jersey's version of S-CHIP, was available to children living in families with incomes up to 350 percent of the poverty line. This was the highest limit in the nation.

p.133 But the money would not last. . . . Martha T. Moore, "States Struggle with Budgets as Money Drips In," USA Today, July 6, 2001; Leslie Berger, "The Rise of the Perma-Temp," NYT, August 4, 2002.

p.134 . . . when Bush was still the governor of Texas . . . Bush, who was in the middle of promoting a $2.7 billion state tax cut, turned down requests to call a special legislative session to pass an S-CHIP program. This refusal meant delaying implementation, because the legislature meets only every other year. And once the legislature did take up the matter, Bush fought to limit eligibility for the program to people below 150 percent of the poverty line. The legislature wanted to raise eligibility to 200 percent—and won. This account is based primarily on my reporting during the presidential campaign of 2000. ("True Colors," *New Republic*, July 12, 1999, p. 15.)

p.134 As a candidate for the presidency in 2000 . . . Robin Toner, "From Social Security to Environment, the Candidates' Positions," *NYT*, November 5, 2000.

p.134 . . . he revived Reagan's old idea of a block grant for Medicaid . . . Robert Pear, "Medicaid Proposal Would Give States More Say on Costs," *NYT*, February 1, 2003; Pear and Edmund L. Andrews, "Bush to Back Off Some Initiatives for Budget Plan," *NYT*, February 1, 2004.

p.134 . . . Jeb Bush, the president's brother, gave the state's . . . John-Thor Dahlburg, "Florida Legislature Passes Governor's Managed-Care Medicaid Bill," *LAT*, December 9, 2005; KFF, "Florida Medicaid Waiver: Key Program Challenges and Issues," Fact Sheet, Washington D.C., December 2005; *Palm Beach* (Florida) *Post*, "Privatizing the Fraud," October 31, 2005.

p.135 . . . Haley Barbour, the former chairman of the Republican Party, tried hard to cut . . . Laura Hipp, "Medicaid, DPS Chiefs Step Down," (Jackson, Miss.) *Clarion-Ledger*, April 27, 2005. This is also based on my reporting about proposed Medicaid cuts in early 2005. ("A Little Trim," *TNR*, February 21, 2005.)

p.135 Bredesen knew a thing or two about health care . . . Shelia Wissner, "Phil Bredesen; Part 1: The Early Years," *Tennessean*, September 29, 2002.

p.135 A lot of what Bredesen proposed to do . . . made sense. . . . Roy Moore, "Back to Plan A: Bredesen Moves to Dissolve TennCare," *Nashville Business Journal*, November 12, 2004; Anita Wadhwani, "U.S. Approves TennCare Limit on Drugs," *Tennessean*, June 9, 2005; Lucas L. Johnson II, "TennCare Director Says Reforms Have Saved Program Millions," Associated Press, November 17, 2005.

p.136 At the tiny North Terrace Family Clinic in Lawrenceburg . . . W. Andrews, P. Burks, and W. Maldonado, interviews with author.

p.137 he became a regular caller on the local radio morning show . . . Ernie Landtroop, interview with author. Landtroop was the morning show's cohost.

p.137 But when the state started talking about cuts in TennCare . . . W. Andrews,

P. Burks, W. Maldonado, Jim Shulman, interviews with author; E. Maldonado medical files; Spragens, "The Faces of TennCare."

CHAPTER 6: CHICAGO

p.141 It was early one Saturday afternoon in the summer of 2001 . . . Marijon Binder, interviews with author and legal briefs in the matter of *M. Binder v. Resurrection Medical Center*; Resurrection Health Care, invoice to M. Binder; Phyllis Pavese, correspondence with author. Pavese was the spokesperson for Resurrection Health Care.

p.143 One reason that Americans have embraced . . . As Larry Gage, president of the National Association of Public Hospitals and Health Systems put it, "America's institutional health safety net has been accomplishing the task of meeting at least the basic needs of the uninsured and underinsured in many areas. It can be argued that this success is the main reason that American politicians have had the luxury of endlessly (and fruitlessly) debating, rather than enacting, universal coverage for the past 50 years." Larry S. Gage, "The Future of Safety-Net Hospitals," in Stuart H. Altman, Uwe E. Reinhardt, and Alexandra E. Shields, *The Future U.S. Healthcare System: Who Will Care for the Poor and Uninsured?* (Chicago, Ill.: Health Administration Press, 1997), pp. 125–126.

p.144 But by the end of the century, skepticism about big government . . . Randall R. Bovbjerg, Jill A. Marsteller, and Frank C. Ullman, *Health Care for the Poor and Uninsured after a Public Hospital's Closure or Conversion* (Washington, D.C.: Urban Institute, September 2000), pp. 1–3.

p.145 A few financially weak or poorly run institutions closed . . . J. Duncan Moore, Jr., "Chasm Grows between Rich and Poor," *Modern Healthcare*, June 7, 1999, p. 34.

p.145 Hospitals streamlined supply chains . . . Julia Flynn Siler and Thane Peterson, "Hospital, Heal Thyself," *BW*, August 27, 1990, p. 66.

p.145 The hospitals of Resurrection Health Care trace their lineage . . . Thomas Neville Bonner, *Medicine in Chicago, 1850–1940* (Urbana and Chicago: University of Illinois Press,), pp. 147–151; and Rosemary Stevens, *In Sickness and Wealth*, 2nd edition (Baltimore, Md.: Johns Hopkins University Press, 1999), pp. 17–18.

p.146 . . . Catholics were the ones who seemed to answer the call most enthusiastically . . . According to one history of Catholics in the United States, "Frequently encountering outrageous anti-Catholic and anti-immigrant prejudice from doctors and hospital administrators, they were often neglected or even denied proper treatment." George C. Stewart, *Marvels of Charity: History of American Sisters and Nuns* (Huntington, Ind.: Our Sunday Vision Publishing Division, 1994), p. 330.

p.146 . . . the Sisters of the Resurrection would not actually establish their hospital . . . Nancy Maes, "Nun's Dream a Reality for Norwood Park," *Chicago Tribune*, October 10, 1984.

p.146 The Holy Sisters of Nazareth, meanwhile, had already established . . . Phyllis Pavese and Rodney Nelson, eds., *A History of Healing, A Future of Care: Saint Mary of Nazareth Hospital Center: Celebrating a Century of Catholic Hospitality* (Flagstaff, Ariz.: Heritage, 1994), pp. 23–52.

p.146 A Polish meatpacker in the city's famous stockyards . . . Chicago Historical Society, "History Files—the Stockyards," viewed online at www.chicagohs.org/history/stock.html; Chicago Public Library, "Chicago in 1900: A Millennium Bibliography," viewed online at www.chipublib.org/004chicago/1900/intro.html.

p.147 In 1897, a survey of patients at yet another Catholic hospital . . . Sarah Gordon, *No Service Too Small: Saint Elizabeth's Hospital* (Chicago, Ill.: Saint Elizabeth's Hospital, 1986), p. 9

p.147. . . nonprofit hospitals were not always eager to take on the indigent . . . Stevens, pp. 10–47 and 108–136. See also Charles E. Rosenberg, *The Care of Strangers: The Rise of America's Hospital System* (Baltimore, Md.: Johns Hopkins University Press, 1987), pp. 238–239.

p.147. . . several of Chicago's hospitals were able to expand . . . Pavese and Nelson.

p.148 But that economic environment disappeared . . . Gary Weiss and Scott Ticer, "A Grim Outlook for Hospitals," *BW*, May 11, 1987, p. 126; Bruce Vladeck, interview with author. (Vladeck is a former director of the Health Care Financing Administration.)

p.148 And they weren't competing just with each other. . . . Paul Starr, *The Social Transformation of American Medicine: The Rise of a Sovereign Profession and the Making of a Vast Industry* (New York: Basic Books, 1982), pp. 428–436.

p.149 The most extreme reaction came from Michael Reese . . . Lucette Lagnado, "Critical Conditions: A Nurse Is Embittered as a Chicago Hospital Fades from Glory," *WSJ*, December 10, 1997. (Originally cited in Stevens.)

p.149 No other hospital in Chicago underwent such a dramatic legal transformation . . . This is probably due to the high number of large teaching hospitals in Illinois, particularly in Chicago. For-profit hospitals never made great inroads in Boston and Philadelphia, either; not coincidentally, both cities are dominated by—some people would say overrun with—large academic medical centers.

p.149 Consolidation allowed hospitals to take advantage . . . James Unland, interview with author. (Unland was president of the Health Capital Group.)

p.150 . . . both Advocate and Resurrection had professional administrators . . . IRS Form 990 filings from Advocate and Resurrection Health Care. Resurrection says the 2002 compensation figure for Toomey is misleading, because in that year he received a retention bonus of more than $1 million. In other years, his compensation was in six figures, not seven. (Source: P. Pavese, correspondence with author.)

p.150 Cardinal Joseph Bernadin of Chicago had long advocated consolidation . . . Bruce Japsen, "Catholic Hospital Firms May Merge," *Chicago Tribune*, June 6, 1999.

p.150 And at Resurrection, officials felt they had fulfilled that mandate. . . . P. Pavese, correspondence with author.

p.151 But in countless ways, big and small . . . Author's visit to Resurrection Medical Center; Michael Zucker, interview with author. (Zucker was an official with the American Federation of State, Municipal, and County Employees who headed up the union's investigations of Resurrection.)

p.152 The story of Marijon Binder's childhood . . . M. Binder, interviews with author.

p.154 Marijon called and arranged an appointment with an official . . . M. Binder, interviews with author; P. Pavese, correspondence with author.

p.154 . . . Marijon did just that, composing a letter in longhand . . . Letter, with attachments, from Marijon Binder to Grabowski and Greene, October 2, 2002.

p.155 Resurrection's attorneys were no more accommodating . . . M. Binder, interviews with author. The descriptions of Marijon's communications with the firm of Grabowski and Greene comes strictly from her recollection and her handwritten notes. The firm, subsequently renamed the Grabowski Law Center, refused numerous requests to discuss the case or any other aspect of its debt collection business. Nor would it contact the attorney from the case, who had left the firm afterward. Efforts to locate him through searches of public records were unsuccessful. Marijon does have a fax receipt to show that the letter was received.

p.155 . . . stories like Marijon's were actually quite common . . . Author's review of records from the Circuit Court of Cook County. Resurrection disputed the significance of the figure, arguing that judges rarely ask for proof when certifying an indigent status application. (Source: P. Pavese, correspondence with author.)

But at least some judges interrogate applicants sufficiently to assess their economic status. And because the legal system does so little to advertise the option, it's likely that many more people would qualify if they knew about it. (Source: A Cook County judge who requested anonymity; A. Alop, interviews with author.)

p.155 ... eventually these lawsuits would start getting attention ... AFSCME Council 31, "Debt Collection Practices at Resurrection Health Care: An Appeal for Compassion," Chicago, 2003. Marijon's story was *not* part of the report. In fact, she says, she was never contacted by union representatives about her experiences with Resurrection Health Care.

p.156 Among the most infamous stories ... Lagnado, "Jeanette White Is Long Dead but Her Hospital Bill Lives On," *WSJ*, March 13, 2003.

p.156 At Provena Covenant, attorneys would ask judges ... Lagnado, "Hospitals Try Extreme Measures to Collect Their Overdue Debts," *WSJ*, October 30, 2003.

p.157 "To put so much silent agony on hapless ... Uwe Reinhardt, interview with author.

p.157 ... Chicago aldermen to Washington power brokers ... Julie Appleby, "Scales Tipping Against Tax-Exempt Hospitals," *USA Today*, August 24, 2004, and Fran Spielman, "Workers Say Resurrection Turns Away Poor," *Chicago Sun-Times*, January 29, 2004.

p.157 Attorneys were getting involved, too ... Daren Fonda et al., "Sick of Hospital Bills," *Time*, September 27, 2004, p. 48.

p.157 Resurrection made no apologies for its firm treatment ... P. Pavese, correspondence with author.

p.158 Hospitals have long approached potential charity cases ... Edna Sproat-Martindale, "Preventing Dispensary Abuse," *Modern Hospital*, January 1935, p. 85. The article was originally cited in Stevens.

p.158 ... more complicated than they seemed ... A case in point was the story of Robin Kemp, one of Resurrection's patients featured in the report by AFSCME.

Kemp, a single mother of two, had recently lost her job when she arrived at the Westlake Hospital emergency room in 2002. She was worried about a painful bump on the side of her face. Doctors at Westlake diagnosed a serious infection and admitted her for several days, during which she was under heavy sedation and high doses of antibiotics. It would take her more than a month to recover fully and several months after that—much of the time spent living with friends, since she couldn't afford an apartment—before she found a new job. But she said she never heard from Resurrection, despite having left a permanent address (a relative's), until, like Marijon, she was served with a summons to appear in court over some $9,000 in outstanding charges. As Kemp remembers it, she immediately called the attorney, hoping to work out an arrangement for payment, and offering to pay $500 right away and $100 in monthly installments after that. The attorney said no; it was too late to negotiate plans. Fearing that a court battle might take away the little money she was now making—or, worse, her kids—she filed for bankruptcy.

Resurrection's version of this story differed from Kemp's, sometimes sharply. As Resurrection tells it, while Kemp was still in the hospital, the staff offered to help her apply for Medicaid coverage and, failing that, financial aid from the hospital. But Kemp declined, saying that she was in the middle of moving and had no address, but promising to get back in touch once she was settled. When she failed to do that, the hospital said, it turned her case over to the lawyers. The hospital also said it had no record of her ever making an offer of payment. "If that offer had been made, it certainly would have been accepted," the hospital said.

In this case and other, similar cases, it was virtually impossible to know whose story was more truthful. (Sources: Robin Lee Kemp, interviews with author; P. Pavese, correspondence with author.)

p.158 Many patients who ended up getting sued by Resurrection . . . M. Binder, interviews with author; P. Pavese, correspondence with author.

p.159 . . . perhaps "cooperation" wasn't nearly as easy as Resurrection made it sound . . . Author's visit to the emergency room at Resurrection Medical Center, June 2004. Resurrection was hardly alone in this regard. See also the Access Project, "Paying for Health Care When You're Uninsured: How Much Support Docs the Safety Net Offer?," January 2003. For more background on EMTALA, the Emergency Medical Treatment and Active Labor Act, see James L. Thorne, "EMTALA: The Basic Requirements, Recent Court Interpretations, and More HCFA Regulations to Come," at the website for the American Academy of Emergency Medicine. Viewed online at www.aem.org/emtala/watch.shtml.

p.159 . . . the handouts from a training session . . ." Resurrection Health Care Presentation on Financial Assistance Proram, April 10, 2002. A spokeswoman for Resurrection said that the presentation was "inconsistent with Resurrection values"–and that, when more senior administrators learned of it, they retrained the hospital staff in ways more in keeping with the hospital's charitable mission. However, the spokeswoman confirmed that some of the policies described in the presentation were in place. (Source: P. Pavese, correspondence with author.)

p.160 Officials at Resurrection insisted that the hospital network . . . Dr. Neil Rosenberg, director of the ICU at Westlake Hospital, went even farther: "I don't think there's another business alive that does close to what we do." (Timothy

McCurry, Neil Rosenberg, interviews with author. McCurry was director of the family practice center at Resurrection Medical Center.)

p.160 Overall, however, the hospital really did seem . . . Judith Graham and Gary Washburn, "Catholic Hospitals Cut Charity Care," *Chicago Tribune*, January 29, 2004; and AFSCME Council 31, *A Failing Mission: The Decline of Charity Care at Resurrection Hospitals*, Chicago, Ill., 2004.

p.160 Certainly, such a reduction would have been entirely consistent . . . Joyce Mann, Glenn A. Melnick, Anil Bamezai, and Jack Zwanziger, "A Profile of Uncompensated Care, 1983–1995," *HA*, July/August 1997, p. 231. See also Raymond J. Baxter and Robert E. Mechanic, "The Status of Local Health Care Safety Nets," *HA*, July/August 1987, p. 7; and Peter J. Cunningham, Joy M. Grossman, Robert F. St. Peter, and Cara S. Lesser, "Managed Care and Physicians' Provision of Charity Care," *JAMA*, Vol. 281, No. 12, March 24–31, 1999. Also: P. Cunningham, interview with author.

p.161 In Chicago, proceedings over medical debt take place . . . Author's visit to the Richard Daley Center; Alan Alop and Katina Cummings, interviews with author.

p.161 Just as Marijon was about to have her turn . . . M. Binder, interviews with author and initial filings in court.

p.161 . . . the attorney told her that she stood little chance . . . M. Binder and H. Brennan Holmes, interviews with author.

p.162 Before managed care, hospitals limited themselves to one set of charges . . . Gerard Anderson, statement before the Committee on House Energy and Commerce Subcommittee on Oversight and Investigations, June 24, 2004. Dr. Anderson was a professor at Johns Hopkins University and a leading expert on hospital pricing.

p.162 Surveys by AFSCME would later find . . . AFSCME Council 31, "A Failing Mission."

p.162 Marijon filed a brief demanding . . . Legal briefs in the matter of *M. Binder v. Resurrection Medical Center*.

p.163 As the hearing began, the judge asked . . . M. Binder, interviews with author. A second eyewitness to the scene, a court officer who requested anonymity, independently confirmed these recollections from Marijon.

p.163 By late 2004, as the talk of political and legal repercussions grew louder . . . Mary Chris Jaklevic, "Attacking Abuse; Lawsuits Plague Not-for-Profits, AHA," *Modern Healthcare*, June 21, 2004, p. 7; and Lagnado, "Hospital Found 'Not Charitable' Loses Its Status as Tax Exempt," *WSJ*, February 19, 2004.

p.164 Advocate Health Care announced . . . Bruce Jaspen, "Hospitals Soften Billing Pain," *Chicago Tribune*, November 2, 2003.

p.164 Resurrection made two conspicuous changes . . . Alop, interview with author.

p.164 It was entirely possible that the critics were right . . . Nancy M. Kane and William H. Wubbenhorst, "Alternative Funding Policies for the Uninsured: Exploring the Value of Hospital Tax Exemption," *Milbank Quarterly*, Vol. 78, No. 2, 2000, pp. 208–209. Also: Kane, interview with author, April 2005.

p.164 As Marijon Binder recounted her saga . . . M. Binder, interviews with author.

CHAPTER 7: LOS ANGELES

p.167 The drive from West 190th Street to Vernon Boulevard . . . Gloria Montenegro, Jose Antonio Montenegro, and Richard Berghendahl, interviews with author; author's visit to Los Angeles.

p.169 . . . the nation's single largest concentration of people without insurance . . . *The State of Health Insurance in California: Findings from the 2001 California Health Interview Survey* (Los Angeles: UCLA Center for Health Policy Research), June 2002. Available at www.healthpolicy.ucla.edu. See also Julie Marquis, "Many in County Remain Uninsured, Survey Finds," *LAT*, October 17, 2000.

pp.169–170 For most of the postwar period, the city was a center of industrial might . . . As Harold Meyerson, one of the area's best-known writers, famously put it, after the cold war the bottom didn't fall out of the Los Angeles economy. The middle did. Harold Meyerson, "California's Progressive Mosaic," *American Prospect*, Vol. 12, No. 11, June 18, 2001; and Jonathan Peterson, "Industrial Blues in the Southland," *LAT*, July 27, 1991.

p.170 Tony was part of this wave. . . . G. Montenegro, J. A. Montenegro, and R. Berghendahl, interviews with author.

p.174 The death rate from cervical cancer in the Montenegros' district . . . In 1999, the age-adjusted mortality rate from cervical cancer in SPA-6 was 5.8; for the United States as a whole it was 2.2, and for rural China it was 4.17. Los Angeles County Department of Public Health, *Key Indicators of Public Health*, 1999/2000; World Health Organization, Worldwide Cancer Mortality Statistics.

p.174 Another set of data showed that . . . For comparisons, see Organization of

Economic Cooperation and Development, *Health at a Glance: OECD Indicators 2005* (Paris, France), 2005, p. 33.

p.175 The two communities had distinct health profiles ... See KFF, "Race, Ethnicity and Medical Care," Washington, D.C., October 1999.

p.175 ... more than one-third of the residents ... Los Angeles County Department of Public Health, *Key Indicators*.

p.175 Eating a proper diet wasn't easy ... See also California Department of Health Services, "Nutrition and Health Barriers Facing California Latinos," Issue Brief, September 2005.

p.176 ... a housekeeper at a local hotel who lived about a mile and a half from the Montenegros ... Mariya Moratoya, interviews with author.

p.176 ... behavior was not exclusively responsible for the health problems ... Hector Becerra, "BP Settles Lawsuit for $81 Million," *LAT*, March 18, 2005; Deborah Schoch, "Study Links Freeways to Asthma Risk," *LAT*, September 21, 2005; and Ben Ehrenreich, "Goo and Gunk," *LA Weekly*, October 19, 2001.

p.176 At Saint John's Well Child Center ... Jim Manja, Richard Morgan, and Paul Giboney, interviews with author. (Manja is president of Saint John's; Morgan is an outreach worker there; Giboney is a physician at the Clinica Monseñor Oscar A. Romero.) For more background on the area's lead poisoning problem, see S.J. Rothenberg, F.A. Williams Jr., S. Delrahim, F. Kahn, M. Lu, M. Manalo, M. Sanchez, D.J. Wooten, "Blood Levels in Children in South Central Los Angeles," *Archives of Environmental Health*, Vol. 51, No. 5, September–October 1995, pp. 383–388.

p.177 Residents who were technically eligible for these programs ... Perhaps nothing better illustrated the confusion than a "decision chart" published by the state, designed to help residents of Los Angeles figure out whether they were eligible for public insurance and—if so—for which program. It was a ten-step flowchart with thirteen boxes, difficult for even a native English-speaking, Ivy League–educated health care journalist to decipher. The sheet, published by the California Department of Public Social Services in July 2002, was available four years later at a URL on the department's website: http://www.ladpss.org/dpss/health_care/healthcare_access_manual_pdffiles/pdf/Health%20Care%20Decisio n%20Chart%202002.pdf.

p.177 The programs also required continual renewals ... During interviews with uninsured residents of Los Angeles County in 2004, 2005, and 2006, these residents repeatedly raised this issue. For a systematic study of so-called churning in Medi-Cal and its effects, see Gerry Fairbrother, *How Much Does Churning in Medi-Cal Cost?* (Woodland Hills: California Endowment), April 2005. As the report notes, churning can end up wasting money, simply by fostering so much additional paperwork and communication.

p.177 ... **used these requirements as a way of trimming program rolls** ... Barbara Lyons, interview with author.

p.177 Although Gloria had gone onto Medi-Cal when she was pregnant ... G. Montenegro and J. A. Montenegro, interviews with author.

p.178 So sometimes Tony got his supplies ... A study published in 2006 confirmed that because uninsured Latinos in Los Angeles were less likely to get routine physical examinations, they were more likely to develop blindness. (Source: Sylvia H. Paz, Rohit Varma, Ronald Klein, Joanne Wu, and Staley Azen, "Noncompliance with Vision Care Guidelines in Latinos with Type 2 Diabetes Mellitus," *Ophthalmology*, Vol. 113, No. 8, August 2006, pp. 1372–1377.)

p.179 Since 1973, its exterior has appeared in the opening credits ... After 1993, the producers of the show replaced the old sequence with a montage of the characters, though a faded image of the hospital building still appeared at the end.

p.179 ... **above the hospital's main entrance is a sculptural group** ... Helen Eastman Martin, *The History of the Los Angeles County Hospital (1878–1968) and the Los Angeles County-University of Southern California Medical Center (1968–1978)* (Los Angeles: University of Southern California Press, 1979), p. 115.

p.179 But in sixty years the burden had grown ... Michael Wilson, correspondence with author; author's visits to the emergency room at Los Angeles County-USC Hospital. (Wilson was a spokesman for the Los Angeles Department of Health Services.)

p.180 One night, not long before Tony had his stroke ... G. Montenegro, interview with author.

p.180 ... **they were worse inside the emergency treatment area** ... Interview with Ed Newton, former chairman of the County-USC Emergency Department, 2006; Brent R. Asplin, "Emergency and Trauma Services in Jeopardy: An Analysis of Proposed Capacity Reductions in the Los Angeles County Department of Health Services," April 18, 2003. For another vivid account of the waiting area at the County ER, see Scott Harris, "Waiting Room," *Salon*, October 21, 1999.

p.181 ... **a group of doctors frustrated by the situation claimed** ... Tracy Weber and Charles Ornstein, "County-USC Doctors Say Delays Fatal," *LAT*, April 23, 2003.

pp.181 Although the subsequent state investigation did not substantiate those charges ... Letter from David Runke, acting CEO of LAC-USC Healthcare Network, to Jacqueline Lincer, California Department of Health Services, "CMS Complaint Validation Survey Plan of Correction," June 30, 2003.

p.181 Medhat Elsadani was a fifty-five-year-old restaurant manager . . . Declaration of Medhat Elsadani, Downey, California, in *Gary Harris v. Los Angeles County Board of Supervisors.* A spokesperson for the county could not confirm or deny stories like Elsadani's. He acknowledged that "long wait times are unfortunately a reality in our public system" and added that "we do our best to triage incoming trauma and ED cases, and people do get shuffled based on acuity, but we strive to meet our mission of care to anyone who walks through our door." (Source: Wilson, interview with author.)

p.182 . . . the system had teetered on the edge of insolvency . . . Stephen Zuckerman and Amy Westpfahl Lutzky, *Medicaid Demonstration Project in Los Angeles County, 1995–2000* (Washington, D.C.: Urban Institute, October 15, 2001). At http://www.urban.org/url.cfm?ID=410295. For this chapter I also drew heavily on my previous reporting. "Ill Treatment," *TNR*, December 2, 2002.

p.183 . . . closing the trauma center at Martin Luther King-Drew Hospital . . . In late 2004, the *Los Angeles Times* ran a five-part series, "The Troubles at King-Drew," by the reporters Mitchell Landsburg, Charles Ornstein, and Tracy Weber, documenting the facility's shoddy care. It laid much of the blame for the hospital's woes—and those of the county health system itself—at the feet of the county supervisors, for mismanagement and obstruction of necessary, if painful, reforms.

p.183 But the county board was also proposing some more dubious changes . . . Asplin, "Emergency and Trauma Services in Jeopardy."

p.183 The county didn't try too hard to sugarcoat the plan . . . The official was John Wallace, a spokesman for the Department of Health Services. Weber and Ornstein, "County-USC Doctors Say Delays Fatal."

p.183 But the county had tried that once before . . . Zuckerman, interviews with author; Zuckerman and Lutzky, *Medicaid Demonstration Project.*

p.184 It was called the Clinica Monseñor Oscar A. Romero . . . P. Giboney and Grace Floustis, interviews with author. (Floustis was the clinic's medical director.)

p.184 . . . according to Tony, the staff at Romero . . . J. A. Montenegro, interview with author.

p.185 Clinica Romero was a federally qualified health center . . . Daniel R. Hawkins, Jr., and Sara Rosenbaum, "The Challenges Facing Health Centers in a Changing Healthcare System," in Stuart H. Altman, Uwe E. Reinhardt, and Alexandra E. Shields, *The Future U.S. Healthcare System: Who Will Care for the Poor and Uninsured?* (Chicago, Ill.: Health Administration Press, 1997), pp. 99–122; Alice Sardell, "Neighborhood Health Centers and Community-Based Care: Federal Policy from 1965 to 1982," *Journal of Public Health Policy*, Vol. 4, No. 4, December 1983, pp. 484–503.

p.185 ... rising support for clinics came at a time of falling support for other safety-net programs ... By one estimate, overall federal spending per each uninsured American actually fell by 8.98 percent from 2001 to 2004. Jack Hadley, Matthew Cravens, Terri Coughlin, and John Holahan, *Federal Spending on the Health Care Safety Net from 2001–2004: Has Spending Kept Pace with the Growth in the Uninsured?* (Executive Summary), Kaiser Commission on Medicaid and the Uninsured (Washington: KFF, November 2005), pp. 1–3. See also Sara Rosenbaum, Peter Shin, and Julie Darnell, "Economic Stress and the Safety Net: A Health Center Update," Kaiser Commission on Medicaid and the Uninsured Issue Paper (Washington: KFF, June 2004), p. 12.

p.186 In Tony's case, the care he got also wasn't enough ... G. Montenegro, J. A. Montenegro, and R. Berghendahl, interviews with author.

CHAPTER 8: DENVER

p.189 Russ Doren loved his wife dearly. . . . Russ Doren and Dan Van Gorp, interviews with author, 2006; R. Doren, letter to Gina Doren's mother, June 18, 1990; R. Doren, letter to G. Doren's mother, July 7, 1990; R. Doren, letter to Cherry Creek Schools, January 14, 1991; Gordon Petersen, M.D., notes in G. Doren's medical file, June 8, 1990; G. Petersen, examination form for Colorado Department of Social Services, July 3, 1990. (Van Gorp was one of Doren's superiors at Littleton High School; Dr. Petersen was Gina's psychiatrist.) The story about Porter's citing financial reasons when it discharged Gina comes strictly from Russ's recollection; officials at Porter declined to comment on the matter, noting that some fifteen years after the event, it was difficult to piece together what had happened. (Source: Rhonda Scholting, interview with author.) Russ's version is certainly plausible, though: discharging patients from inpatient psychiatric care was a common practice by the 1990s, as discussed later in the chapter.

p.192 Care for people with psychiatric disease has always lagged ... Steven M. Sharfstein, Anne M. Stoline, and Lorrin M. Koran, "Mental Health Services," in Anthony R. Kovner and Steven Jonas, eds., *Health Care Delivery in the United States*, 7th ed. (New York: Springer, 2002), pp. 246–247; Sharfstein, interview with author.

p.193 One obstacle to obtaining better resources and oversight ... Shaftstein, Stoline, and Howard H. Goldman, "Psychiatric Care and Health Insurance Reform," *American Journal of Psychiatry*, 150:1, January 1993, p. 8.

p.193 That started to change in the early twentieth century ... Sharfstein et al., "Mental Health Services," pp. 246–247.

p.193 A wave of deinstitutionalization ensued . . . Thomas G. McGuire, "Financing and Demand for Mental Health Services," *Journal of Human Resources*, Vol. 16, No. 4, Autumn 1981, p. 503.

p.193 President Kennedy substantially increased funding . . . President John F. Kennedy, Special Message to the Congress on Mental Illness and Mental Retardation, February 5, 1963; Sharfstein, interview with author.

p.194 The early Blue Cross plans of the 1930s hadn't covered psychiatric illness . . . McGuire, p. 505.

p.194 . . . by the late 1950s . . . Herman Miles Somers and Anne Ramsey Somers, *Doctors, Patients, and Health Insurance: The Organization and Financing of Medical Care* (Washington: Brookings Institution, 1961), Sharfstein, interview with author. For more history, see U.S. Department of Health and Human Services, *Mental Health: A Report of the Surgeon General* (Washington, D.C., 1999).

p.194 More serious psychiatric disorders . . . Sharfstein and Stoline, "Reform Issues for Insuring Mental Health Care," *HA*, Fall 1992, pp. 85–86.

p.195 . . . 99 percent of corporate group health policies had acquired . . . "A New Market for Hospital Chains," *BW*, April 11, 1983, p. 124.

p.195 . . . people had become far more willing to talk about their mental afflictions . . . Although it is widely accepted that attitudes toward mental illness have grown more tolerant over time, there is not much empirical evidence to document the change—apparently because nobody conducted serious polls on the subject until the late 1980s. Andrew B. Borinstein, "Public Attitudes towards Persons with Mental Illness," *HA*, Fall 1992, p. 196.

p.195 . . . businesses that invested in mental health treatment were also acting out of self-interest . . . Robert P. Frederick, "Alcoholics Get Help at Work," *NYT*, December 31, 1981; "More Help for Emotionally Troubled Employees," *BW*, March 12, 1979, p. 97; "A New Market for Hospital Chains," *BW*, April 11, 1983, p. 124.

p.195 Scrutiny immediately turned to psychiatry . . . John Kaas, "Enough Is Enough, Youth Psychiatry Critics Say," *Chicago Tribune*, May 31, 1989.

p.196 . . . after for-profit facilities had targeted such clients aggressively . . . Milt Freudenheim, "The Squeeze on Psychiatric Chains," *NYT*, October 26, 1991.

p.196 "Psychiatric care is obviously out of control" Freudenheim, "Mental Health Costs Soaring," *NYT*, October 7, 1986. The analyst was Kenneth Abramowitz.

p.196 Looking back on the early days of their relationship . . . R. Doren, interviews with author, 2006; G. Doren, letter to her grandparents, August 20, 1986;

G. Doren, letter to her mother, August 20, 1989; G. Doren, letter to her mother, July 30, 1990. Gina's mother declined to comment on the story of her daughter.

p.202 **Research showed that managed care approved hospital stays for psychiatry** ... Stuart Steers, "Mind Games: The Squeeze in Mental-Health Care Is Choking Patients and Providers Alike," *Denver Westword*, May 24, 2001. The research was by Thomas Wickizer and Daniel Lessler.

p.202 ... **a new line of business emerged to take advantage of the new opportunity to make a profit** ... Rhoda Donkin, "The New Mental Health Watchdogs: Can They Deliver?" *Business and Health*, February 1, 1989, p. 16. For more on the clinical impact of these new businesses, see Wayne A. Ray et al., "Effect of a Mental Health 'Carve Out' Program on the Continuity of Antipsychotic Therapy," *NEJM*, Vol. 348, No. 19, May 8, 2003, pp. 1885–1894.

p.202 **Such savings would prove typical of the new environment** ... John Iglehart, "Managed Care and Mental Health," *NEJM*, Vol. 334, No. 2, January 11, 1996, p. 133; Richard G. Frank et al, "Mental Health Policy and Psychotropic Drugs," *Milbank Quarterly*, Vol. 83, No. 2, pp. 271-298; HayGroup, "Health Care Plan Design and Cost Trends" (Arlington, Va.: April 1999).

p.203 ... **psychiatrists complained loudly** ... Diane Levick, "Insurers Cut Back on Mental Health Coverage," *LAT*, October 22, 1989.

p.204 ... **the medical director for Aetna's psychiatric management unit admitted** ... Freudenheim, "The Squeeze on Psychiatric Chains," *NYT*, October 26, 1991.

p.204 ... **a private corporate benefit manager was even blunter** ... Daniel Forbes, "The Draconian Cuts in Mental Health: Employee Benefits," *Business Month*, Vol. 130, September 1987, p. 41.

p.204 **Gina was proof that even the most severe cases** ... R. Doren, interviews with author; G. Doren, correspondence.

p.205 ... **the Cherry Creek School District had switched policies** ... R. Doren, letter to District Insurance Committee, Cherry Creek Schools, November 23, 1987; R. Doren, letter to Carroll McNeill, Insurance Committee Chairperson, Cherry Creek Schools, April 11, 1988.

p.205 **And the Dorens exhausted that limit after the first hospitalization** ... R. Doren, interviews with author.

p.206 **The mounting debt weighed heavily** ... G. Doren, note left on September 20, 1990. Gina also referred to her worries about mounting debt in an angry letter she wrote to her mother during the summer of 1990: "On top of all of this our insurance ran out months ago, and we are in such financial difficulty that I may have to go to a state hospital. We have even considered bankruptcy." G. Doren, letter to her mother, July 23, 1990.

p.206 . . . **shifted more responsibility for psychiatric care to the government** . . . Sharfstein, interview with author.

p.206 . . . **Colorado owned and operated two mental hospitals** . . . Bill Scanlon, "Fort Logan's Hospital Could Be Bleak," *Rocky Mountain News*, April 22, 2003; Alan M. Kraft, "The Fort Logan Mental Health Center," *Milbank Memorial Fund Quarterly*, Vol. 44, No. 1, Part 2, January 1966, pp. 19–20.

p.207 Gina was wary of the transfer . . . R. Doren, interviews with author, 2006; R. Doren, letter to Gina's mother, June 18, 1990; R. Doren, letter to Gina's mother, July 7, 1990.

p.208 Depression and suicidal tendencies are common . . . S. Sharfstein, interview with author.

p.208 One day, when Russ arrived at the hospital . . . R. Doren, interviews with author; R. Doren, letter to Gina's mother, July 7, 1990.

p.208 But then she decided to drive to Cheyenne overnight . . . R. Doren, letter to Dr. Petersen, September 26, 1990.

p.208 . . . **Gina had gotten in touch with her former family physician** . . . In a note Gina left after taking the medication, she wrote a message to her stepfather: "Thanks for the pills." (G. Doren, note left on September 27, 1990.)

p.209 Russ returned from work to find Gina . . . R. Doren, interviews with author; Aurora Police Department Report, case Number 90-39718, incident on September 27, 1990, filed October 11, 1990.

p.209 Russ had started to wonder what Gina's true goal had been . . . R. Doren, interviews with author, 2006; letter from R. Doren to G. Doren (addressed posthumously), October 30, 1990.

p.211 Years later, the Colorado legislature would consider passing . . . David Algeo, "Mental Illness Gets Parity," *Denver Post*, December 28, 1997.

p.211 In 1996, the federal government enacted a version of parity . . . Deborah Sontag, "When Politics Is Personal," *New York Times Magazine*, September 15, 2002, p. 90.

p.212 Not long after the laws had passed, it became clear . . . M. Audrey Burnam and José J. Escarce, "Equity in Managed Care for Mental Disorders," *HA*, Vol. 18, No. 5, pp. 22–31; David Mechanic and Donna D. McAlpine, "Mission Unfulfilled: Potholes on the Road to Mental Health Parity," *HA*, Vol. 18, No. 5, pp. 7–21; Alan L. Otten, "Mental Health Parity: What Can It Accomplish in a Market," (New York: Milbank Memorial Fund), June 1988.

p.212 . . . reformers had hoped that the initial parity law would be a stepping-stone . . . Rebecca A. Markway, "Wellstone Legislation Languishing," *St. Paul* (Minn.) *Pioneer Press*, June 15, 2004.

p.212 He took Gina's death hard . . . R. Doren and D. Van Gorp, interviews with author.

p.213 But in some ways, Cherry Creek proved a tougher sell . . . Robert D. Tschirki, superintendent of Cherry Creek Schools, letter to R. Doren, January 31, 1991.

p.213 It reminded Russ of the reaction he'd gotten before . . . R. Doren, interviews with author, 2006.

CONCLUSION: WASHINGTON

p.215 . . . public opinion about health care in America remains as ambivalent . . . In a poll from 1993, when the cause of health care reform was at its peak, a (bare) plurality of 43 percent agreed that the "uninsured are able to get the care they need from doctors and hospitals." In 1999, the figure was up to 57 percent. Over the years, pollsters have also asked respondents whether they thought the uninsured were predominantly employed or unemployed. (Most of the uninsured have jobs.) Majorities consistently answer "unemployed." (Robert J. Blendon, John T. Young, and Catherine M. DesRoches, "The Uninsured, the Working Uninsured, and the Public," *HA*, Vol. 18, No. 6, pp. 203–211; Blendon and John M. Benson, "Americans' Views on Health Policy: A Fifty-Year Historical Perspective," *HA*, Vol. 20, No. 2, March/April 2003, pp. 33–46.)

pp.217–218 . . . the figure will be up to 56 million . . . According to this projection, the overall growth of the population would account for only one-third of the increase. Overall, 27.8 percent of the working non-elderly population would have no coverage. Todd Gilmer and Richard Kronick, "It's the Premiums, Stupid: Projections of the Uninsured through 2013," *HA*, Web Exclusive, April 5, 2005, pp. W5-143 through W5-151. (Note that they use a model that correctly predicted the uninsured rate of 2002, on the basis of past data.)

The question of exactly how many people are uninsured has become a subject of some controversy. But among experts, a rough consensus exists that, at any one time, between 40 and 45 million people have no health insurance. For more on the various estimates, and why they differ, see Economic Research Initiative on the Uninsured, "How Many Are Uninsured? Different Data Offer Different Dimensions," Research Highlight No. 6, August 2004; and KFF, "Who Are the Uninsured? A Consistent Profile Across National Surveys," Issue Paper, August 2006.

p.220 . . . the creation of Health Savings Accounts . . . This is based primarily on my past reporting about HSAs. ("Crash Course," *TNR*, November 7, 2005, 18–23). For a recent study on the impact of high cost-sharing, see D.P. Goldman et al., "Varying Pharmacy Benefits with Clinical Status: The Case of Cholesterol-lowering Therapy," *The American Journal of Managed Care*, Vol. 12, No. 1, January 2006, pp. 21–28.

p.224 . . . the vast majority of doctors . . . Peter J. Cunningham, Andrea Staiti, and Paul B. Ginsburg, "Physician Acceptance of New Medicare Patients Stabilizes in 2004–2005" (Washington, D.C.: HSC, January 2006). More than 96 percent of physicians responding to the survey said they were accepting at least some new Medicare patients, and nearly 73 percent said they were accepting all new Medicare patients. Both figures were higher, albeit only slightly, than the corresponding figures for physicians accepting patients with private insurance.

p.224 . . . the elderly have far more positive feelings about Medicare . . . Karen Davis, Cathy Schoen, Michelle Doty, and Katie Tenney, "Medicare Versus Private Insurance: Rhetoric and Reality," *HA*, web exclusive, October 9, 2002. Doctors and hospitals complain, legitimately, that Medicare's coding system for billing is maddeningly complicated. But any modern health insurance system will require complicated reporting. Many doctors and hospitals also don't like the fact that Medicare sets fees—and that the fees are lower than those paid by private insurance. But Medicare also tends to pay faster than private insurance, with less interferences in the clinical process. And, for providers, the reward for putting up with Medicare's drawbacks has been a massive infusion of cash into the health care system. A world without Medicare would be a world in which a large portion of the elderly would have no health insurance, as was the case in the years before Medicare came into existence. They would either get no health care at all or depend on charity. Either way, it would ultimately mean less money for doctors and hospitals.

p.225 Yet there is precious little evidence that our extra spending makes us healthier. World Health Organization, *The World Health Report 2000* (Geneva, Switzerland), 2000; Organisation for Economic Cooperation and Development, *Health at a Glance, OECD Health Indicators, 2005* (Paris, France); Gerard Anderson and Peter Sotir Hussey, "Comparing Health System Performance in OECD Countries," *HA*, Vol. 20, No. 3, May/June 2001, pp. 219–232; Anderson, Uwe E. Reinhardt, Peter S. Hussey, and Varduhi Petrosyan, "It's the Prices, Stupid: Why the United States Is So Different from Other Countries," *HA*, Vol. 22, No. 3, May/June 2003, pp. 89–105. Taken one at a time, none of these measures is definitive. Some measures reflect international differences in reporting standards. The United States, Canada, and the Scandinavian countries have a different standard for live births from most of continental Europe; that affects the statistics on infant mortality, tending to make the United States look slightly worse than it actually is. Even with the more carefully calibrated figures, it's hard to tell how much a number reflects the health care system and how much it reflects other factors, such as poverty rates and diet. Japan, for example, has conspicuously high mortality rates for some cancers, but most experts think that those reflect genes and diet—not health care. Still, when the United States ranks merely average on so many measures, it's hard to make a credible argument that we have the "world's greatest health care system," as proponents frequently claim, or that universal health care leads to worse health.

p.225 You'll hear instead about rationing and waiting lines . . . See, for example, Clive Crook, "Poison Pill," *The Atlantic*, April 2006, and David Gratzer, "Where Would You Rather Be Sick?" *WSJ*, June 15, 2006. For a more detailed version of these arguments, see Michael Cannon and Michael Tanner, *Healthy Competition: What's Holding Back Health Care and How to Free It* (Washington: Cato Institute), 2005. One of the more seemingly persuasive arguments these critics make is that the death rates from several cancers is higher in Canada than in the U.S.—proof, these critics say, that Canada doesn't treat disease as well. But those death rates are on the number of people diagnosed with the disease—and it turns out that the U.S. diagnoses these diseases much more frequently than Canada. Overall, the proportion of Americans who actually die from these cancers is roughly the same as the proportion of Canadians who die from them. (In fact, the proportion varies very little across the developed world with just a handful of exceptions.) What does this mean? One possibility is that, because the U.S. system screens more aggressively for these diseases, it turns up many more cases that might otherwise go undetected—in other words, slow-moving tumors that would not actually kill a person until long after something else had caused that person's death. The data do support the contention that the U.S. health care system is more aggressive than Canada's. And sometimes more intense treatment is better. But it's not clear that's true here. (Sources: Anderson and Hussey, "Multinational Comparisons of Health Systems Data," Commonwealth Fund, October 2000, pp. 17–20; Hussey, Anderson, Robin Osborn, Colin Feek, Vivienne McLaughlin, John Millar, and Arnold Epstein, "How Does the Quality of Care Compare in Five Countries?" *HA*, Vol. 23, No. 3, May/June 2004, pp. 89–99.

p.226 The stories about Canada are wildly exaggerated. . . . Steven J. Katz, Diana Verrilli, and Morris L. Barer, "Canadians' Use of U.S. Medical Services," *HA*, Vol. 17, No. 1, January–February 1998, pp. 225–235.

p.226 The British spend just 7 percent of their national wealth on health care . . . Reinhardt et al. *HA*. It's important to remember that spending levels sometimes reflect underlying cultural preferences. That seems to be the case with Britain. The classic study of this is Lynn Payer, *Medicine and Culture* (New York, Penguin, 1988).

p.226 A perfect example is Japan. . . . OECD, "OECD Health Data 2006: How Does Japan Compare," viewed online at www.oecd.org/dataoecd/30/19/36959131. pdf.

p.226 Still, the best showcase for what universal health care can achieve may be France. . . . In France, even elective medical procedures are available more or less on demand. The exception is at the major teaching hospitals: if you want to see the top specialist in a particular medical field in Paris, you may have to wait for an appointment. But that's no different from the situation in the United States, where you may have to wait for an appointment if you want to see the top specialist at Johns Hopkins. In fact, contrary to the myths about universal health care, the French actually get more medical care than we do. The French have more hospital beds and see their physicians more than their U.S. counterparts. Thomas C. Buchmueller and Agnes Couffinhal, "Private Health Insurance in France," OECD Health Working Papers No. 12 (Paris: OECD, 2004); Michael K. Gusmano, Victor G. Rodwin, and Daniel Weisz, "A New Way to Compare Health Systems: Avoidable Hospital Conditions in Manhattan and Paris," *HA*, Vol. 25, No. 2, 2006, pp. 510–520; Jean-Pierre Poullier and Simone Sandier, "France," *JHPPL*, Vol. 25, No. 5, October 2000, pp. 899–905; Rodwin and Sandier, "Health Care under French National Health Insurance," *HA*, Fall 1993, pp. 111–131; J. Hurst, Gaetan LaFortune, Rodwin, interviews with author. (Hurst and LaFortune are researchers at the OECD.)

p.229 . . . "problems will not be solved with a nationalized health care system . . . President Bush, State of the Union Address, Washington, D.C., 2003.

AFTERWORD

p.233 Graeme Frost was a bespectacled . . . Matthew Hay Brown, "The Education of the Frost Family," *Baltimore Sun*, October 10, 2007; David M. Herszenhorn, "Capitol Feud: A 12-Year-Old Is the Fodder," *NYT*, October 10, 2007; Michelle Malkin, "Graeme Frost and the perils of Democrat poster child abuse," Michelle Malkin website, October 8, 2007; retrieved at michellemalkin.com/2007/10/08/ graeme-frost-and-the-perils-of-democrat-poster-child-abuse/; Samantha Sault, "Daily Blog Buzz: S-chip's Real Recipients," *Weekly Standard* website, October 7, 2007, retrieved at www.weeklystandard.com/weblogs/TWSRP/2007/10/ daily_blog_buzz_1.asp; Mark Steyn, "Brother, can you spare a CHIP?", *National Review Online*, October 7, 2007, retrieved at corner.nationalreview.com/ post/?q=MDAzYiY5OWVkMmQxZTJmNTZlNDNjZTlhOGU3NjNlZDA=; Bob Vineyard, "$45,000 and No Insurance," *Insureblog*, September 28, 2007, retrieved at insureblog.blogspot.com/2007/09/45000-and-no-insurance.html; Vineyard, David Lemmon, correspondence with author. (Lemmon was a spokesperson for FamiliesUSA, a liberal health care advocacy group.) After the extensive media coverage of the first few weeks, the Frosts declined all interview requests, including mine.

p.236 It was true that the Frosts weren't destitute . . . U.S. Census Bureau, *Income, Poverty, and Health Insurance Coverage in the United States: 2006* (Washington, D.C., 2007), pp. 18-26; National Center for Health Statistics, *Health, United States, 2007* (Hyattsville, MD, 2007), pp. 56-58. As the report notes, "The burden of out-of-pocket health care expenses is greatest for poor and uninsured people. But some higher-income families with health insurance who have catastrophic illnesses or high out-of-pocket expenditures for noncovered services may devote a substantial portion of their income to medical care, or to health insurance premiums, or both."

p.236 The talk had started in the late fall of 2006 . . . This is based primarily on my reporting about Wyden's bill and the 2008 presidential campaign, including interviews with the candidates, their advisers, and outside experts. (See, in particular, "What's the One Thing Business and the Left Have in Common?" *New York Times Magazine*, April 1, 2007; "Cautious Candidate, Cautious Plan," *TNR* online, May 30, 2007; "Medical Miracle," *TNR*, September 10, 2007; "The Mulligan," *TNR*, October 8, 2007.)

p.240 . . . Republican presidential candidate Rudy Giuliani introduced . . . This is based primarily on my campaign reporting. (See, in particular, "Giuliani's False Promise," *TNR* online, August 2, 2007.) Ezra Klein, "A Man With a (non-) Plan, *American Prospect* online, August 2, 2007; Paul Krugman, "The Substance Thing," *NYT*, August 6, 2007.

p.240 As public health experts were quick to note . . . Gemma Gatta, et al, "Toward a Comparison of Survival in American and European Cancer Patients," *Cancer*, Vol. 89, No. 4, August 15, 2000, pp. 893-900; Bengt Jonsson and Nils Wilking, "A Global Comparison Regarding Patient Access to Cancer Drugs," *Annals of Oncology*, Vol. 18, Supplement 3, April 2007, pp. 1112-1117; Arduino Verdecchia, et al, "Recent cancer survival in Europe: a 2000-02 period analysis of Eurocare-4 data," *The Lancet Oncology*, August 21, 2007, published online at oncology.thelancet.org; The Commonwealth Fund, "Statement by the Commonwealth Fund on Use of Prostate Cancer Statistics," October 30, 2007, published online at www.commonwealthfund.org/newsroom/newsroom_show.htm?doc_id=568333; Krugman, "Prostates and Prejudices," *NYT*, November 2, 2007; Wilking, Merrill Goozner, interviews with author. (See also "Creative Destruction," *TNR*, November 19, 2007; "Health Care Like the Europeans Do It," *TNR* Online, April 10, 2007.)

p.241 Other new studies painted an even dimmer picture . . . Carlos Angrisano, et al, "Accounting for the Cost of Health Care in the United States," McKinsey Global Institute, January 2007; Karen Davis, et al, "Mirror, Mirror on the Wall: An International Update on the Comparative Performance of American Health Care," Commonwealth Fund, New York, May 2007; Ramesh Ponnuru, "Everything You Always Wanted to Know about Health Care—But were afraid to ask," *National Review*, September 24, 2007. For more background, see Bianca Frogner and Gerard Anderson, "Multinational Comparisons of Health Systems Data, 2005," Commonwealth Fund publication no. 825 (New York, NY, April 2006); David Cutler, *Your Money or Your Life* (New York: Oxford University Press, 2006), pp. 10-31; Ellen Nolte and Martin McKee, "Measuring the health of nations: analysis of mortality amenable to health care," *British Medical Journal*, Vol. 327, November 15, 2003, pp 1129-1136.

p.241 They said the government estimates on the uninsured . . . David Gratzer, "First, Do No Harm," *Forbes*, February 12, 2007; Greg Mankiw, "Beyond Those Health Care Numbers," *NYT*, November 4, 2007.

p.241 While these arguments held some element of truth . . . Economic Research Initiative on the Uninsured (ERIU), "How Many Are Uninsured? Different Data Offer Different Dimensions," Research Highlight No. 6, August 2004; ERIU, "A Revolving Door: How Individuals Move In and Out of Coverage," Research Highlight No. 1, October 2002; KFF, "Who Are the Uninsured? A Consistent Profile Across National Surveys," Issue Paper, August 2006; Michael Perry and Julia Paradise, "Enrolling Children in Medicaid and S-CHIP: Insights from Focus Groups with Low-Income Parents," KFF May 2007; Karl Kronebusch and Brian Elbel, "Simplifying Children's Medicaid and SCHIP," *HA*, Vol. 23, No. 3, May/June 2004, pp. 233-246. This is also based on my interviews.

ACKNOWLEDGMENTS

This book would not have been possible without the participation of the people who appear in it as characters—and who spent long hours, sometimes days, sharing intimate, frequently painful details about their lives. They did so for no compensation, other than the satisfaction of knowing their experiences might enlighten other people. For that I express both my gratitude and admiration. And because only a handful of these people actually appear in this book, I'd like to acknowledge a few others who were particularly generous with their time: Helen Cothron, Ed and Colleen Gangswich, Suzanne Gibbons, George Iwerks, Robin Lee Kemp, Margaret Loncar, Donnie Magnum, Juan Nieto, Carol Niewinski, Aixa Reyes, Beatriz Reyes, Rose Schaffer, Patricia Simmons, and Sherri Walton.

When I set out to write this book, I frequently told people that narrative reporting would inevitably complicate—and, ultimately, strengthen—my ideas about health care policy, gleaned from years of political reporting. I'm not sure whether I fully believed it at the time, but I do now. And yet trying to write a book this way created one obvious challenge: Finding people to interview. Local newspaper articles helped me pinpoint some subjects, including two of the book's main characters: Russ Doren, whose story I first read in an article by David Algeo in the *Denver Post*; and the Maldonados, who were mentioned in a story by John Spragens in the *Nashville Scene*. But for the most part I was dependent upon intermediaries at health clinics, community outreach groups, and other such organizations who introduced me to their own contacts and, in many cases, helped arrange interviews. Among them were Michael Campbell, Grace Floustis, Lark Galloway-Gilliam, Paul Giboney, Ida Hellinger, Angela Hyde, Mandy Johnson, Terence Long, Jim Manja, Patrick McCabe, Roland Polencia, David Roman, Nancy Rothman, Sylvia Ruiz, Brady Russell, and Evonne Tisdale. A few of these intermediaries went even further, spending days of their time introducing me to people and showing me around their communities, then returning my countless phone calls and e-mails as I tried to nail down reporting facts. I owe them not only gratitude, but gas money and, probably, a meal or two as well: in Chicago, Alan Alop, Katina Cummings, and Meg Lewis-Sidme; in Philadelphia, Patricia Gerrity; in Lawrence County, Patsy Burks; in Los Angeles, Carl Coan, Danny Feingold, Simmi Gandhi, Loretta Jones, Richard Morgan, and the organizers of UNITE/HERE Local 11.

To fill in my knowledge of local history, I had the assistance of librarians at the Sioux Falls Public Library, the Los Angeles Public Library, and the Chicago Historical Society; to fill in my knowledge on everything else, I had the help of the staff at the University of Michigan libraries. Elizabeth Brennan, at the ACLU of Southern California, made available to me that organization's considerable files about the Los Angeles County health care system. I'm also grateful to Leigh Eckmair of the Gilbertsville Free Library, who introduced me both to the village's history and many of its inhabitants.

Deciphering the academic literature on health care is no small task—particularly for somebody who slept through a few too many classes as an undergraduate. Fortunately for me, I've managed to get an informal graduate education from a de facto committee of experts that includes Robert Blendon, Gary Claxton, David Cutler, Robert Cunningham Jr., Paul Fronstin, Nancy Kane, and Uwe Reinhardt. Two of the country's best health care journalists, Karl Stark of the *Philadelphia Inquirer* and Charles Ornstein of the *Los Angeles Times*, selflessly shared their considerable knowledge, as did two industry insiders: James Unland and Bruce Vladeck. Two other experts who I consulted frequently deserve special mention, as I am proud to call them not only teachers but also good friends: Jonathan Gruber of M.I.T., and Howard Markel of the University of Michigan.

A few people did me the added favor of reading portions of my book in draft form: Patricia Neuman of the Kaiser Family Foundation lent her expertise on Medicare; her colleagues Barbara Lyons and Samantha Artigua did the same for Medicaid. Georgetown's Mila Kofman helped me track down the victims of insurance fraud—and then taught me everything I needed to know about the individual insurance market. Steven Sharfstein walked me through the modern history of psychiatry; his son, Joshua, helped me in countless ways. Jacob Hacker of Yale somehow found time to help me even as he was finishing his own book. Jonathan Oberlander of the University of North Carolina and Theod re Marmor of Yale basically adopted me as a part-time student, sharing their wisdom on health care—and, in Marmor's case, college football. Matthew Holt and Joe Paduda, two industry consultants whom I met through their respective websites, gave me invaluable comments, as did Ezra Klein, a frighteningly knowledgeable young writer at the *American Prospect*. Consultant John Larew, plus attorney Joseph Palmore and physician Tara Palmore also lent their expertise to chapters of mine. Needless to say, any errors that remain in the text are mine and mine alone.

One benefit of living in a college town is the plentiful supply of smart, energetic students to help with research. I was fortunate to find a few such students: Megan McMillan, Zachary Peskowitz, Sara Schmitt, and Alexandra J. Sloan. Kate Steadman, another precocious young expert, helped me with research, as well.

In retrospect, my professional interest in health care policy traces back to my first permanent job in journalism, as an assistant editor for the then-fledgling *American Prospect*. Health care policy was a passionate interest for both of the magazine's editors, Bob Kuttner and Paul Starr, long before I came along. That was particularly true for Starr, whose 1986 book, *The Social Transformation of American Medicine*, remains the essential text on the history of American health care. Only later would I come to appreciate fully the value of the policy knowledge they imparted and the intellectual discipline they promoted.

Still, it was not until I came to my present professional home, the *New Republic*, that I actually wrote about health care. A succession of editors—Charles Lane, Peter Beinart, and Franklin Foer—have nurtured this interest, giving me the room to explore the topic seriously in a way few other publications would. I'm grateful to Martin Peretz, for providing me with this journalistic home, as well as to Leon Wieseltier, for encouraging me to be an intellectual grown-up. Over the years, I have benefited time and again from editors who have selflessly improved my prose—particularly Nurith Aizenman, Sarah Blustain, Jeremy Kahn, Christopher Orr, and Peter Scoblic. No less indispensable has been the all-too-frequently invisible work of fact-checkers and copy editors who saved me countless embarrassments—in particular, Ben Adler, Kara Baskin, Alexander Belenky, Josh Benson, Ruth Franklin, Marisa Katz, Joe Landau, Adam Kushner, Anne O'Donnell, Clay Risen, Ben Soskis, and Sacha Zimmerman. *TNR* interns Eve Fairbanks and Elspeth Reeves moonlighted as my freelance research assistants in the book's final weeks. Thanks, also, to Bruce Steinke and Henry Riggs, whose patience for my endless last-minute corrections on articles related to this book defies any reasonable expectations of common courtesy.

Among my friends are some enormously talented fellow journalists, several of whom graciously read portions of this book and improved it immeasurably: Christopher Caldwell, Dan Costello, Kate Marsh, Noam Scheiber, Gordon Silverstein, and Jason Zengerle. Jonathan Chait and David Grann read chapters—and dealt with my neurotic phone calls—even though they had their own books to write. Jessica Dorman did me the Herculean favor of reading every single page, lending the artful touch of a historical novelist; so did Carol Mainville, who, although a physicist by training, turns out to have a superb editing eye. John Judis read everything and gave me his characteristically blunt feedback, even as he was doing the same for several colleagues. I mean no disrespect to his own, considerable body of work when I say his greatest contribution to journalism may be on the writers like myself he's mentored over the years.

Chapter seven of this book is adapted from an article that I wrote for the *New York Times Magazine*. My thanks to editors Catherine St. Louis and Alex Star for the assignment, and to Sarah Smith for the fact-checking. The other indispensable ingredient in this effort was a media fellowship provided by the Henry J. Kaiser Foundation. Aside from financial support, education, and exposure to health care stories, it provided me with an introduction to the indefatigable Penny Duckham, who has become a fairy godmother to dozens of journalists like me around the world.

Whatever the final verdict on Hillary Clinton's 1994 health care plan, reporting this book has convinced me the former First Lady was right on another issue: it really does take a village to raise a family. When it came to raising mine, I was fortunate to have these people in my village: Lana Donnelly, Renata Mendes, Serena Simmons, Bibi Lorber, Jeff Micale and Marybeth Damm, Jim Hoeffner and Priti Shah, and the staff at UMCCWF. My village frequently extended to family in Massachusetts and Florida, sources of crucial emotional support and babysitting. Thanks also to the staff at Sweetwaters on Ashley Street, for their generous supply of caffeine, wireless access, and good cheer.

Young writers, in particular, need not only inspiration but also an assist or two. I was fortunate to get both, many years ago, from Linda Healey and her late husband, J. Anthony Lukas. The assist was a referral to Kathy Robbins, who despite a client list full of luminaries managed not to forget about me and, along with her army of capable assistants, got me writing when the time was right. Kathy also matched me with Tim Duggan, an editor whose reputation for enthusiasm, congeniality, and talent are entirely deserved. Like Kathy, Tim believed in me—and this project—on the many occasions I did not. Plus he somehow managed not to laugh aloud every time I promised to meet a newly postponed deadline.

My father, Leon Cohn, M.D., read the entire book for medical accuracy. Still, I doubt that he and my mother, Elaine, realize how much they did to make this book possible—or how grateful I am to both of them.

My older son, Tommy, is slightly older than this book project, but neither he nor his younger brother, Peter, can remember a time when I wasn't working on it. They gave up far too many stories, expeditions, and baseball games so that I could have "just a few more minutes of work time." I intend to have made up for that by the time this is actually published, but I would still like to thank them—not just for the patience they showed but also for the inspiration they provided me every day, by reminding me through their mere existence that sometimes all can be right with the world.

As for my wife, Amy, I fear that any acknowledgment here will be an insultingly small gesture given what she has done for me these last few years. Balancing career and family is never easy in a couple with two overachievers. It must be near impossible when one of them is as congenitally disorganized and professionally obsessive as yours truly. She put up with me in part, I think, because she thinks this book is important—which is testimony to her deeply held sense of social justice. Whatever the reason, I am grateful for her indulgence and for a few other things, too, like her bright blue eyes and wickedly smart sense of humor. Mostly, though, I am grateful that she decided to attend an alumni get-together in Boston on a late summer night many years ago. I have cherished every moment since.

INDEX